Academic Entrepreneurship

Gary E. Harman

Academic Entrepreneurship

Springer

Gary E. Harman
Professor Emeritus
Cornell University
Geneva, NY, USA

ISBN 978-3-031-06820-1 ISBN 978-3-031-06821-8 (eBook)
https://doi.org/10.1007/978-3-031-06821-8

This Springer imprint is published by the registered company Springer Nature Switzerland AG
The registered company address is: Gewerbestrasse 11, 6330 Cham, Switzerland

First and foremost, the author wishes to thank his wife Jean, who is the love of his life. Without her, none of this would have been possible. She helped the author even when this was uncomfortable.

The author also wishes to thank individuals who played an important role in his career. One such person is Professor Ralph (Tex) Baker of Colorado State University who gave the author a job when he was an undergraduate, which set him on the path of his career. Another is Ilan Chet, Professor at the Hebrew University of Jerusalem. He has been a close friend, and he and the author partnered in grants that provided the necessary funding, without which his career could not have happened. He also wishes to thank Drs. Matteo Lorito and Sheri Woo. This husband-and-wife team have been partners throughout his career. He also wishes to thank Professor Norman Uphoff, who has been a cooperator since the author's retirement.

He also wishes to thank his company co-founders, including Rus Howard, Tom Stasz, Tom Bourne, and Terry Spittler. Dan Custis, CEO of Advanced Biological Marketing, who led this company to financial success. There also have been excellent Board of Director members, including Professor David Call, Carl Fribrolin, and John Hicks.

The author also wishes to thank students and postdocs, including Marc Blume, Tom Stasz, Ellen Chirco, Akrofi Djetpror,Judith Hourgh, Chaur Tseun Lo, Joyathi Bolar, Fatemeh Mastouri, Maros Feris, Jonthan Hubbard, Izhad Hadar, Wei-Lang Chao, Eric Nelson, Alex Sivan, Xixian Jin.Shipeng Deang, and Bruno Donzelli. He also wishes to commend Glenda Nash, Pat Neilsen, Kristen de la Fuenta, and Robert Patrick—these technicians were a very strong part of the team. He also wishes to thank Gerald Marx, Morrill Vittum, and Thomas Björkman, who are Cornell colleagues.

Three persons at the Cornell patent office have been very helpful, including Walter Haeussler, Dick Cahoon, and Alan Paau.

Finally, the author wishes to thank the Cornell Biotechnology Program and the US-Israel Binational Science Foundation. These two organizations are the only ones that would provide funding for the translational research, and without which his career could not have succeeded.

Introduction and Preface

The author of this book has been engaged in developing technologies and establishing start-up companies based on his university technology since 1993. He is aware that the route to commercialization from a university base may be difficult and that there is little guidance or even encouragement. This book is written to provide insights into experiences in this. An alternative title might be "Thing that I wish I had known when I first began start-up company activities."

There are attitudes and behaviors that are common to those beginning entrepreneurial activities. This framework was very well expressed by Bob Noyce and Bill Hewitt, which is etched on the Tech Museum of San Jose, CA. These two are among the leaders that led to the establishment of the Silicon Valley. They express the spirit and the institutional requirements that lead to the creation of new things that are at the heart of the entrepreneurial ideal, as follows:

> OPTIMISM IS AN ESSENTIAL INGREDIENT FOR INNOVATION. HOW ELSE CAN THE INDIVIDUAL WELCOME CHANGE OVER SECURITY, ADVENTURE OVER STAYING IN SAFE PLACES?
>
> AN ABIDING CURIOSITY AND AN INSATIBLE DESIRE TO LEARN HOW AND WHY THINGS WORK ARE THE HALLMARKS OF INNOVATION...CREATIVITY IS NUTURED BY BEING RECEPTIVE AND ENCOURAGING.

The first is a call to entrepreneurs everywhere and expresses the excitement and adventure that lead to the creation of something new, which never existed before. The second also relates to the necessary curiosity and unquenchable desire to learn, which is the hallmark of successful entrepreneurs in and out of academe. It also refers to the essentiality of institutional receptivity and encouragement that is, in many cases, lacking in the academic environment, to the detriment of both the scientists involved and to the university itself. It also addresses the essentiality of a willingness to venture outside safety, which is a necessary attribute of entrepreneurship.

This book required focus since the paradigms and practices involved in different types of intellectual endeavors required for different fields differ markedly. Furthermore, universities differ in their approaches and limitations to entrepreneurship, and it is not possible to cover their entire breadth. Therefore, the author chose to focus on biology and agriculture since this is his area of expertise. However, entrepreneurship in these fields has not been comprehensively addressed, in contrast to fields such as computer and information technologies. It is hoped that examples from biology and agriculture will be illustrative of needs and challenges in the important area of academic entrepreneurship across diverse fields and institutions.

Focus on a restricted area of endeavor is essential. Attempting to cover all universities and disciplines would result in an unwieldly book. A similar tight focus was the subject of another book on this topic (Shane, 2004), which focused on the Massachusetts Institute of Technology. University policies between different universities are dissimilar in numerous ways. What is accepted and the norm at one university may not be appropriate in another.

The scope of this book must be built upon a definition of entrepreneurship. This word means different things to different people. An excellent description is from the University of Rochester in upstate New York. They state that "we understand entrepreneurship to mean the transformation of an idea to an enterprise that creates value—economic, social, cultural, or intellectual. More than a discrete set of business skills or practices, entrepreneurship is a calling that can be pursed in many realms of experiences and achievement. A core value of American culture, entrepreneurship uniquely combines the visionary and the pragmatic. It requires both individual initiative and knowledge, and through awareness of markets and attention to the needs of others. Entrepreneurship is a way of thinking, an approach to problems, an attitude of mind, and even a trait of character. It is a science and an art; entrepreneurship is a primary way in which a free society grows and improves not only its economy, but its cultural and social life as well." (from https://www.sas.rochester.edu/entrepreneurship.html).

Thus, the concept of entrepreneurship is not only about starting new companies. Starting a new business is one possible outcome, but it is just a means to an end. Drucker (1993) provides a broad definition of entrepreneurship, that is, entrepreneurs "create something different; they change or transmute values." Of course, universities are exactly in that arena—they certainly create something different and transmute value, so by that definition they ought to be highly entrepreneurial.

As will be described in this book, in some cases the only possible way to achieve an entrepreneurial goal is through start-ups or licensing to an existing company, and in other cases, goals can be reached without a corporate intermediate. Examples and practices with and without a commercial entity will be described, and it depends on the nature of the technology involved. Highly entrepreneurial programs are initiated to improve a product or good that already exists. This type of effort will be described in this book as *adaptive*. In this case, the desired end product can be sold or distributed through already-existing channels. However, in other cases, the desired end product is a totally new product and will be described in this book as *disruptive*. A disruptive product has to be invented, refined, manufactured, and sold or distributed to a skeptical public. If a product is adaptive, a new commercial entity may be unnecessary, but if it is adaptive then company formation is likely to be required. Adaptive products or services are more amenable to academic practices, because conflict of interest issues are avoided, and the transition from lab to practice are less. In this book, examples of adaptive and disruptive technologies are provided.

One surprising fact ascertained in researching this work is that there are no books that have addressed the tightrope and the maze that university-based entrepreneurs must walk through if they are to be successful. Given that there have been about 3400 US university-based start-up companies (Anonymous, 2009), this is surprising. This means that there are number of like-minded university faculty that are creating companies and that many of them are groping in a fog, both in respect to good business practice and operating under poorly defined rules governing activities of such faculty in universities. This book hopes to lighten that fog.

One excellent book does address academic entrepreneurship (Shane, 2004), but it was written by a business author based on interviews and evaluation of information from the Technical Transfer Office (TTOs), inventors, and the universities (primarily MIT) themselves. A very important divergence occurs because, in my experience, the view of commercialization of technologies developed by university scientists is greatly different between the scientists and business writers. Unfortunately, there are no books from one who has "walked the walk and talked the talk." The difference in viewpoint and approach of a business writer, a university, a TTO, and a practitioner is going to be very different.

The vast majority of universities and other research institutions have active (and frequently expensive) TTOs, which have the mission of providing intellectual property protection to university patents and securing development of commercial opportunities. However, other administrative units in the universities are many times discouraging, or at best, indifferent, to the activities of faculty who are trying to bring their discoveries to market and to create public good and wealth. On the one hand, university administrators would like their discoveries to be commercialized and to gain the royalties and other financial gains from these activities. However, they frequently lack experience and so do not have a good appreciation of how a strongly proactive technology commercialization program can be managed, including issues such as conflict of interest and commitment. The book will also describe reasons why university discoveries are frequently left to languish from a commercial perspective, even if major academic publication and recognition are attained.

Return could be augmented by changes in policies that encourage and reward faculty inventors who create new economic opportunities for the taxpayers, students, granting agencies, and others who support them. University start-ups also have a high success (survival) rate. The survival rate of academic start-ups is around 70%, while for start-ups in general the 10-year survival rate is only about 30% (Shane, 2004, 2010).

In spite of these partial successes, it appears to the author that many of the policies of at least Cornell are designed to impede (although not intentionally) success in the generation of royalties and other benefits that ought to come to the University and its scientists. This book will outline concrete steps that can, if implemented, increase substantially the entrepreneurial efforts and rewards to both scientists and the institution.

Personal Advantages

When the author first began to consider entrepreneurship, and especially to become involved with commercial companies, some Cornell administrators advised against this course of action.

They felt that his career was going well, and that the commercial involvement would detract and impede my scientific career. Looking back, this advice was simply incorrect. The author's academic and commercial career had been immeasurably enhanced by blending of academic science and commercial activities. There are three evidences of this. First, many of the initial observations that led to the majority of the basic academic studies that the author developed were first observed in commercial trials. Having access to hundreds of trials conducted by commercial partners/cooperators provides a huge advantage. Second, the commercial success achieved had contributed to validation of the basic studies, which could not have been accomplished in any other way. In most cases, this assisted in both publications and grant proposals, but, of course, there were the situations where clearly reviewers negatively reacted to the commercial entities. Finally, the author's reputation worldwide has been hugely enhanced by the blend of solid science and commercial success. The use of beneficial microorganisms in commercial agriculture is largely attributed to this combination, as is frequently stated to him: That could not have occurred without the commercial successes. The success of the current commercial entity, Advanced Biological Marketing, has permitted the author to continue his career as its chief scientific officer after his retirement from Cornell. This is permitting a capstone to his career of enhanced financial well-being and to create a legacy that will continue through the university discoveries that are frequently left to languish from a commercial perspective, even if major academic publications and recognition is attained.

Purpose and Target Audiences

The purpose of this book is, first and foremost, to challenge colleges and universities with large research programs to be more proactive than they currently are in ensuring that their technologies reach the marketplace. The universities themselves are the primary audience.

A second audience for the book are potential university entrepreneurs, and effort has been made in this book to describe to them the issues and rewards involved, including how university cultures affect, or may affect, success or failure of the enterprise.

A third audience is students, many of whom are fascinated by the concept and process of creating value from university technologies and in creating companies based upon them. However, there simply are few guidelines or road maps on this topic, and students frequently become discouraged. They usually, especially at the graduate student level, conduct research in their chosen field and do not become familiar with business concepts or have any idea how technology transfer works, primarily because their major professors are unfamiliar and/or uncomfortable with this process.

Finally, the author hopes to describe to others, including persons who may be interested in participating in a university entrepreneurial program, potential investors and interested persons who simply would like to see universities take a more active role in creating wealth for this nation and the world.

The Need for Academic Entrepreneurship

Drucker (1993) points out that "nothing could be as risky as optimizing resources ...where the proper course is innovation." Universities today are in perpetual states of financial difficulties and are doing their best to optimize their resources, which can be defined as moving assets to accomplish the same goals as before. To obtain their operating funds, the preferred options are increasing student tuition, overhead on grants, and increasing endowments, largely from alumni. Moreover, to an increasing degree, universities are becoming more regimented, are increasing administrative support staff, and increasing unit sizes. This increasing level of regimentation and increased unit size run counter to what the author's views as critical—creating greater entrepreneurial opportunities and grass-roots efforts from faculty. Across at least the USA, the view of most faculty is that the environment for creativity is becoming more limited and that the general atmosphere is stultifying. Smaller groups with similar foci generally are more creative. One of Cornell's greatest strengths used to be "creative chaos," which was an atmosphere in which unusual and creative ideas could easily emerge. This strength seems to be waning and increasing bureaucracy taking its place.

A potential method to innovate, in Drucker's sense, is to embrace entrepreneurship. In terms of this book, this equates primarily to creating an environment where technological advances are transformed into useful products and services that create wealth in the world, the nation, and in the communities in which they reside. In many, but not all, cases, this requires improved methods, especially university cultures that encourage patenting and then commercialization of the patents they generate.

Academic entrepreneurship can be a rewarding and joyful passage that also is likely to result in late night awakenings to the cold realities of cash flow, salaries, and personal debt and responsibilities. One aim of this book is to reduce the number of 3 am vigils on the part of the academic business start-up partners.

One obvious advantage for universities in entrepreneurship is Ito increase royalties. University income from royalties is not small, but it is the author's contention that it is a fraction of what it could be if entrepreneurship was effectively managed, rewarded, and administered in a faculty-friendly way. This requires, in most cases, improved practices, including fostering of university cultures that encourage entrepreneurship, and where appropriate, patenting, and then commercialization of the patents they generate. If they do this, then the following will occur:

- The universities will generate significantly more royalty revenue from inventions by faculty.
- They will create jobs and wealth around the world. Colleges and universities create jobs and wealth, but typically they do not quantify jobs or wealth, nor do they provide rewards for faculty that are effective in this regard.
- They will increase funding opportunities for their faculty, which is a critical measure of success. The old "publish or perish" paradigm has been replaced by "get funding (preferably with high overhead rates) or perish" because there are no other ways to generate successful research programs.
- They will provide opportunities for faculty to benefit through sharing royalties. The first item above needs amplification. Universities and other research institutions have significant TTOs. The TTOs provide valuable services and are responsible for administering technology transfer including managing patents and licenses. These generate significant revenues for universities and, by extension, create significant increases in employment and wealth. In the US, the Association of University Technology Managers (AUTM) records and chronicles activities. In 2009, the total license income to all universities was $2.1 billion. The majority of this income was from licenses on patents so the income to universities was large. This income is not equally distributed; of the 179 universities that responded, the top 25 accounted for 83% of the total income. This certainly must translate into a substantial amount of wealth creation and jobs, but that data on job and wealth creation is usually not collected by most universities. This is surprising—legislators frequently seek information on benefits to their constituents. Specific data on job and wealth creation is, or should be, necessary data provided by university researchers. If very specific job creation numbers were available to university administrators and could be conveyed to legislators, this would go far in defending public research budgets. This would be especially true if universities created a climate fostering entrepreneurship, job and wealth creation.

Moreover, if we assume that, at a maximum, the royalty income to universities accounts for 3% of the total economic returns to companies licensing the technologies, then the total economic benefit from the licenses is $70 billion annually; in the year 2000, the benefit was estimated at $33.5 billion (Cohen, 2000). This value derived from AUTM probably is an underestimate because this organization only records and uses data on intellectual property that has an initial license fee of $1000 or more, and many licenses do not meet this requirement, even though they may become highly successful (communication from Alan Paau, Director, Cornell Center for Technology, Enterprise, and Commercialization). In the author's area of upstate NY, just two institutions (Cornell University, including the medical campus in NYC and the University of Rochester) received about $70 million in royalty and related income according to their most recent annual reports, which, using the same 3% rate values, would result in $2.3 billion in economic activity. This is a significant return, but one that the universities with which I am familiar do not publicize nor use to their advantage.

The Structure of This Book

Chapter 1 deals with the critical issue of the university culture. Some universities are more welcoming of for-profit enterprises, while others are less so. Adaptive technologies fit the universities more easily than adaptive ones. Indeed, some faculty members regard any type of for-profit activities with disdain. The author of this book has heard that he "doesn't fit the academic mold." If a faculty member or staff wishes to reach into commercialization, then it is very important to understand how this will be viewed by his colleagues. If anyone wants to proceed with a commercial endeavor, it is likely to be possible, but in some universities, this may be a lonely proceeding. In some universities, such at MIT, private start-ups are encouraged. This is possible because in the surrounding community there are many investors willing to fund them. In these situations, faculty must have their own lab and other facilities. Other areas, including the author's, such funding is not readily available, so start-ups are more difficult.

Chapter 2 deals with the reasons why a university faculty member might want to become highly entrepreneurial, whether with or without involvement of a commercial entity. A primary reason that is repeated again and again (e.g., (Shane, 2004)) is that this is many faculty wish to see their work in practice and benefiting the general public. Many of us conduct basic research and publish, and this is both satisfying and fulfilling. However, if the research ends with only publication in scholarly journals, regardless of the impact factor of the journal, this may not be enough for many of us. The conversion of a basic scientific principle or discovery to a widely useful product that contributes to innovation and to the well-being of society can be rewarding both financially and emotionally. This is a primary reason why the author started companies, and as we will see, it is also true for many others. Such motivations are not confined to academics; Kawasaki (2004) describes entrepreneurship as creating meaning above all, which has little to do with money, power, or prestige.

Entrepreneurial activities are strongly influenced by the product that is desired. In some cases, the end product is an improvement either in the product itself or in novel manufacturing/production systems. In an academic setting, it frequently is possible to accomplish the goal of the improved product without establishment of direct commercial entity. In this case, the products are provided through the standard marketing chain. This chapter will explore the adaptive and disruptive technologies just mentioned. In many cases, publishing the results in scientific journals is not enough—especially if the academic founder strongly desires that the products, services, or goods need to be translated into materials that increase economic well-being and jobs or the societal benefits that the technology would bestow. The choice of developing an adaptive or a disruptive technology is largely driven more by the nature of the product than the choices of the inventor/entrepreneurial scientist.

There are two "valleys of death" in the creation of useful applications of university technologies. The first is the conversion of scientific discovery to a useful product, and the second, which applies more to disruptive than adaptive technologies, is the development, production, and marketing of the product of the program to a commercially viable commodity.

Chapter 3 contains case histories of several entrepreneurial programs in biology. These range from genetic and molecular approaches to biological innovation, including methods to introduce DNA into cells (the gene gun) or universal methods to alter organisms through CRISPR, and two methods to enhance rice production, albeit through very different approaches. The McCouch rice program aims to enhance production of this very important crop through plant breeding methods, while the other aims to enhance environmental sustainability and increase production through integrated management systems. Another effort seeks to enhance markets for broccoli in the NE through a combination of improved varieties and marketing systems. The final example is the author's own long-term efforts to improve agricultural sustainability, resistance to biotic and abiotic stresses by use of symbiotic fungi in the genus *Trichoderma.* All of these projects have been successful and provide benefits to plant productivity and economic benefits.

Chapter 4 deals with a critically important component of academic entrepreneurship, which is conflict of interest (COI). COI is an increasingly difficult issue, and academic entrepreneurs must successfully deal with it. COI issues are usually more difficult for scientists developing disruptive technologies than adaptive ones, since disruptive technologies usually require a commercial entity, funding from the commercial entity for product development (not allowed at some institutions), and a personal profit motive for the academic scientist involved. The reasons why COI policies are required and necessary are discussed. Nearly all institutions recognize that COI is frequently present in academic programs, and the policy is to manage but not eliminate COI. Reporting of potential COI is required by most academic scientists, and if COI issues are revealed, then there are faculty COI committees (fC0I) that must respond to each case. Depending on the degree of COI discovered, different COI resolutions will be prescribed for different situations. This is all necessary and reasonable, and this chapter will provide specific situations where COI is present, and even describe some that should be absolutely prohibited for a full-time faculty member. There are both conflicts of financial interest and conflicts of commitment, and both must be managed.

Chapter 5 deals with the critical area of agreements, contracts, regulatory affairs, and patents. These legal documents are very important components of the entrepreneurial efforts. These must be understood and pursued correctly. If this is not done, then the efforts are likely to fail, and there are unpleasant consequences for improperly managing these legal documents.

Finally, *Chap. 6* deals with the formation of companies from an academic base, including suggestions as to how to secure the funding to pursue programs. Funding is critical for any academic but is especially important for anyone wishing to commercialize products. Salaries, manufacturing facilities, marketing, and the other necessities of companies must be accomplished. Sources of funding and pitfalls in securing this will be considered.

References

Anonymous, (2009). AUTM U.S. Licensing Activity Survey FY 2009. AUTM (www.autm.net), Deerfield, IL.

Cohen, W. (2000). Taking care of business. ASEE Prism Online, 1-5.

Drucker, P. F. (1993). Innovation and Entrepreneurship. HarperBusiness, New York.

Kawasaki, G. (2004). The Art of the Start. Penguin Books, London.

Shane, S. (2004). Academic Academic Entrepreneurship. University Spinoffs and Wealth Creation. Edward Elgar, Northhampton, MA.

Shane, S. (2010). The Illusions of Entrepreneurship: The Costly Myths That Entrepreneurs, Investors, and Policy Makers Live By.

Contents

Chapter 1
University Culture

Nothing is as important to the success or failure of faculty-based start-ups or other entrepreneurial activities as the culture or the university in question. In Chap. 3 are examples of different types of entrepreneurial projects. Some are adaptive and create useful solutions for existing industries, and these usually do not involve new companies. In other cases, the technologies are disruptive and require entirely new products, production systems, and marketing. Some of these generate royalty incomes while others do not. Regardless, the university culture strongly affects all of these; the university culture may be among the most important factors determining success or failure. In some institutions, there is a stigma associated with any activity where the profit motive is a part of the equation. In other institutions, start-ups and other entrepreneurial activities such as those described in Chap. 2 are welcomed and encouraged. This makes all the difference.

In institutions where entrepreneurial activities are discouraged (either intentionally or more likely unintentionally), the atmosphere of disapproval tends to pervade both faculty and administration. In welcoming institutions, the reverse is true. Moreover, there may be a great difference between activities involving for-profit companies and not-for-profit organizations. If the institution does not actively encourage entrepreneurs, the structure frequently is negative, and the impression of official statements frequently is punitive, or that is the perception. The tone is set by the administration. Compare the following two statements:

> "Faculty and staff members must take special care to separate their university responsibilities for research and education from their engagement with and commitments to external entities (companies), including Cornell-associated startups, in which they hold a financial interest. Most conflicts of financial interest and commitment arising from faculty/staff involvement with a Cornell-associated startup can be successfully managed. The goal of the University and the faculty/staff should be to work collaboratively to develop an effective management plan that is transparent and protects the integrity of Cornell research, ensures compliance with applicable regulations and institutional policies, protects students' academic interests, fosters an open academic environment, and ensures the primary professional

G. E. Harman, *Academic Entrepreneurship*,
https://doi.org/10.1007/978-3-031-06821-8_1

> commitment of full-time faculty and staff to the university" (from the policy statement of the Cornell fCOI committee)
>
> "Every week, we look forward to seeing faculty members and students coming through our office doors with inventions they want to turn into startup companies. Some are serial entrepreneurs. Some are brand new to the process. Regardless of where you are on the learning curve, we're here to help make your startup happen" from Lita Nelsen, Director, MIT Technology Licensing Office.

Both statements are accurate assessments and internally correct. The first statement suggests a great deal of red tape and suggests consequences to any improper activity. Institutional support other than handling the regulations does not appear to be available. The second statement is welcoming and indicates that there is institutional help and support for crossing and managing the chasms that exist between academic research and entrepreneurship. The first is a substantially negative message while the second is a welcoming and supportive one. The first deals with faculty conflict of interest solely, and both MIT and Cornell have strong COI policies, and, in fact, the MIT position is stricter than that at Cornell that at MIT, faculty cannot receive grants for their research program from companies in which they have an equity position.

The difference is much greater than just style. There exists at MIT a strong tradition, supported by local VC and other investment groups, of faculty who start their own companies; a recent book suggested that about half of the MIT faculty are engaged in start-ups (Shane, 2004). At Cornell entrepreneurial faculty are relatively rare.

If one accepts the principles that are implicit in this book, which include:

- Universities are among the primary developers of intellectual capital in most of at least the developed world.
- A great deal of the discoveries and inventions at universities can be crafted into useful products or services that benefit society, both in terms of the betterment of society and the important financial advantages provided.
- Universities should have and accept the obligation to create jobs and foster economic growth for the taxpayers and others who have supported their activities.
- For reasons already described, many universities do a poor job of fostering useful entrepreneurship.
- Many universities do not take steps that can optimize income in the form of royalties and other benefits.

Acceptance or rejection of these principles have real consequences both for the institutions and for the society that provides, in one way or another, the funding that permits their existence and for the institutions themselves.

Thus, some universities, and even divisions within universities, have much better records in academic entrepreneurship than others. In the case of poorly performing institutions, there are a few relatively simple changes that can improve their situation, assuming that the institution wants to make changes. Other changes are more difficult and expensive. There are two general categories of changes that will be discussed here, as follows:

1. Provide reward including especially one that contribute to academic advancement and promotion for entrepreneurial activities
2. Provide effective programs to mentor and assist entrepreneurial activities including start-ups. These programs will need to be tailored to the institution and the communities in which it is located—for example, the entrepreneurial program at MIT would translate very poorly to upstate NY. MIT has access to substantial venture and other capital and so many or most start-ups exist in private laboratories funded by investors. This asset is largely lacking in upstate NY, and so a different model is needed.

The University Reward Structure

The university reward structure needs to enhance and reward the risk-taking and extra effort required to launch entrepreneurial activities. Such activities include start-ups, but as will be documented in broccoli and rice (Chap. 3), other types of entrepreneurial activities can be highly advantageous to the university and the community in which it exists.

Faculty and staff at academic institutions generally have well-defined reward systems that they can meet to obtain success and promotions. These typically include, for research appointments, publications in peer-reviewed journals, which a premium is given for those with high impacts; invitations to speak and/or publish papers from professional societies; highly visible program efforts recognized by various types of media; the success of advanced degree students that they have mentored; and generally becoming highly recognized in their field of endeavor. Persons with teaching programs are recognized by their classroom contact hours, by reviews of students and others, and by the successes of their students, whether undergraduate or graduate. Persons with extension appointments again are recognized by the quality of information that they provide, by the contacts they make, and by the success with which their programs result in techniques that benefit growers or other consumers of their information. These requirements, just like those for conflict of resolution requirements, are stringent. Thus, faculty are accustomed to meeting such requirements.

There are objective criteria and defined quantitative criteria that can be provided to measure success of entrepreneurial activities if institutions are willing to adopt these. They can be added to the classical research, teaching, and extension reward criteria. If universities accept the criteria of need stated earlier in this chapter and those of first principles at the start of this book, then these ought to be implemented. Such reward components are listed and described below.

Rewards Based on Patented Technologies

1. Inventors receive a portion of the royalties paid on inventions. This is a standard practice at universities, but the author has yet to meet an inventor who views this reward as significant or of great importance. It is at best nice to have but secondary to the goal of creating a valuable product or service. The reasons for this are several: (1) most inventors did not invent a technology primarily to make money; they did this because it was an important part of their research program and because it was critical if their technology was to become societally useful, either economically or because of the intrinsic value of their discoveries, and (2) the money derived is, in the majority of cases, not very large and probably not worth the extra time and effort that patent filing requires as opposed to other activities, such as grant and Ms writing, student mentoring, etc., that are more valuable, and from the university and professional standing perspective, that are more rewarding.
2. Inventors receive professional and university standing for successful patents. University faculty that the inventor knows receive little or no promotional or professional credit for successful patents. This ought to change, but there are different levels of success and they ought not to be judged equally.

   a. The first level is writing and filing a patent (this needs to be through the university patent office as noted in Chap. 2). Such filings need to be approved by the patent office but this is not a rigorous review and no guarantee of success. Patent filings ought to be listed but not taken as a strong indication of success—in terms of comparison with publications, it is probably equivalent to publication in a non-refereed journal. It is, though, a first step.
   b. Another level is that the patent actually issues. However, in the USA, at least in the author's area of endeavor, even first reviews of patents by the US or other patent offices required two or more years. By the time all the reviewers' objections are satisfied, with the long time periods required for their response, times to actual patent issue are frequently more than 5 years from filing. This long time period makes issued patents not a very useful criterion for success.
   c. The most useful patents or patent applications are the ones that are licensed by a commercial or other entity who will practice the invention. Patents that are licensed means that the licensing entity sees value in the invention and anticipates making a profit or at least a useful product in the case of nonprofit organizations. This is an important measure of past success. Licensed patents are items that universities or other public institutions need to consider strongly in their review of faculty performance. In terms of equivalence to scientific publications, these ought to be considered on a par with publication in a quality reviewed journal. In the author's experience at Cornell, data on numbers or success in licensed patents was not even collected by university administrators, which translates directly into an implied statement that licensed patents are not considered valuable at the institution.

d. An even more telling criterion of patent success is the amount of royalties that an individual academic scientist provides to the university. Royalties, when divided by the royalty rate, give an exact measure of the economic benefit that a particular invention has to society. Moreover, the royalties, if significantly large, can provide a direct economic benefit to the institution. In the Preface, the author stated that the amount of royalties generated by licensed inventions is small relative to the potential. This certainly is true and universities and other institutions can benefit if they institute appropriate systems that reward entrepreneurship.

Rewards Based on Job Creation, Economic Benefits to the Community, and Enhanced Competitiveness of Existing Industries

Some entrepreneurial activities do not result in patents and therefore the components of reward listed above do not apply. However, they still may be very valuable in terms of economic and societal benefits. In some cases, these do not necessarily result in publications or other components of reward. Moreover, even for technologies that include patents, the value and potential reward to the individual faculty member are underestimated. An alternative and additional reward ought to be based on the economic and social benefits conferred by the technology. Here are some quantifiable parameters that can be used, but they require more effort by the university to document.

Job and economic benefits are an important component to entrepreneurial and conventional academic pursuits. Moreover, legislators and other stakeholders in the institution frequently request such data to justify investment in the institution. Clearly, legislators and other public and private officials will be inclined to look favorably on institutions that are providing substantially increased employment and economic benefits for the nation, state, or region that they represent. If the institution can readily provide such information, especially solid and quantifiable data, then they are likely to be much better off in the budgeting process. The more jobs and the more economic or societal benefits that can be documented by the institution, the better. Moreover, these jobs and benefits ought to be outside the institution. The author has heard comments from university administrators indicating that job creation outside the institution is not important, and they are creating jobs in the institution and that is sufficient. There are two fallacies to this argument: (1) the jobs created within the institution result primarily from taxes or other sources, including private gifts, contracts, and grants, so these jobs are diversion of public wealth to academic purposes. This is not improper, but it does not result in the creation of new wealth or really jobs, since if that money was not provided to the institution, it might have resulted in more jobs in the society and companies, and (2) these jobs do not result in increased economic benefit in the same way that sales of products would do. Thus, universities need to have methods of quantifying the values of their technologies in direct improvement of the economics and social value of their efforts.

Entrepreneurial scientists and the institutions they represent need to partner to quantify the economic benefits of their entrepreneurial activities. Here are some of the methods that can be used for quantification.

1. Job creation: In many cases, entrepreneurial scientists are intimately associated with the companies created or made possible by their inventions or developments. Again, these may be patented or not, but when scientists are engaged directly with the companies or industries they are benefitting, then they can provide reliable data on both the numbers of new jobs and the quality of the jobs they create. If such an intimate relationship is not the norm, then for patented technologies, the transfer office ought to require that the numbers of jobs, and their location, be provided to the office. This, in most cases, is sensitive business information and so the license agreement ought to (a) require the data and (b) agree in writing that the information is confidential and will not be publicly divulged other than with a court order. In other cases, it will not be possible to demand such data as part of a licensing agreement. In this case, the university and/or investigator needs to discuss with the industry or company to determine the number of jobs and their quality and this needs to be reported to the university (on a confidential basis if necessary). The institution can benefit mightily from such information because it provides a direct and solid base for reporting to legislators, granting agencies or others that provide funding. If such information is obtained on a confidential basis, it can be reported in aggregate and can provide highly useful information. When job creation occurs as a consequence of a faculty member or team, the investigator and/or team needs to receive credit for promotion, raises, and other benefits of university employment.
2. Wealth creation: If the technology or program results in sales of products or services, then increased wealth is created in the community or world. If the technology is patented, then the royalty stream will provide a direct measure of this value. If no patents are involved, then, just as described above, the university and the investigator or team need to work together to discover this value. The university will benefit directly from improved relationships with taxpayers and investors. The faculty member or team ought to benefit in the reward structure of the institution.
3. In other cases, the benefit is to world society in general. The direct benefit of the rice program described in Chap. 3 is to provide resources that can enhance the worldwide research effort, enhance productivity of this important staple crop, and provide both food a wealth to countries and individuals that need it most. The calculus of benefit on an exact economic basis is difficult in this case, but probably is still quantifiable, for example, on improvements in rice productivity and the economic benefit of this improvement.

Institutional Entrepreneurial Support

If a scientist or a team of scientists are considering establishing a start-up company within a university structure or if an investor is considering funding of such a group, then they need to look at the university structure to determine whether or not such efforts will be welcomed and encouraged. Included in this examination are a number of factors, some of which will be obvious from the preceding discussion. These include:

1. A determination of the university or institutions conflict of interest policy and whether either the scientist(s) or investors find the rules and regulations acceptable.
   a. Perhaps most importantly, is the management C01 transparent to the investigator? Is it possible for a potential entrepreneur to meet with a representative of the fCOI to determine criteria up-front? If there is no possibility of the potential for discussion of issues with the committee or individuals involved in COI, the author's recommendation is either not to engage in activities covered by these groups or to change employment to a more enlightened institution.
2. A discussion with the technology transfer office (TTO) to determine policies and practices likely to be encountered (assuming patenting is likely to be involved), including:
3. Most TTOs have a policy in which the investigator submits a relatively brief invention disclosure. This includes such topics as the area of patenting and, importantly, whether or not there has been any prior public disclosure. It also will request information regarding any companies that (a) have funded the invention,
4. Whether this funding provided any rights of first refusal to IP resulting from the invention, and (c) whether the inventor knows of any companies that may be interested in licensing of the invention.
5. In addition, patents are expensive and so the TTO will want to determine whether the patent costs will be reimbursed for the invention. This is an important consideration, particularly if a new start-up will be responsible for this cost burden. A discussion should be initiated with the TTO to determine the cost burden to the licensee. The author is aware of institutions that expect large sums up-front for a license, an expectation of recovery of patent costs, and quite high royalty rates (in most cases, 2–3% royalties are appropriate in the agricultural sector, although they may be higher in other fields). As discussed in Chap. 2, inventions made by university scientists frequently have a great deal that must be done in order to create useful products in the marketplace. If the TTO appears to have unduly high expectations especially in up-front fees, it is unlikely that a license will be possible. It is important to know going in what expectations will be, and whether the TTO, on behalf of the institution, will be willing to, at best, break even financially initially in terms of direct outlays (not including the TTOs salary) with the

expectation/hope that the patent will be profitable. Usually TTOs will have boiler plate agreements that the investigator can view. Anyone new to this area ought to discuss what is reasonable in terms of TTO fees with someone knowledgeable in the field of discussion.

6. Talk with other scientists in the institution regarding their experiences with the TTO. Items that are large negatives are long turnaround times, unreasonable financial demands are discussed in the previous paragraph, and difficulties of the TTO in finding and dealing with licensees are common complaints. On the other hand, there may be scientists with quite positive experiences with the TTO; in general, the author's experience with this office at Cornell has been very positive. A former head of Cornell's HO helped locate a potential CEO that was a very useful first step. With a new start-up, it is useful to discuss any perceived deficiencies with the TTO offices or officers. It also needs to be emphasized, as discussed in the last chapter, that the patent needs to be owned by the university or other academic institution where the invention occurred. The reasons for this were discussed there, but the concept expressed in the AAUP report (AAUP, 2012) states that an individual faculty owns the intellectual property they develop while an employee of the institution needs to be resisted. The reasons why faculty wish to own the IP they develop is rooted in large part on at least the perception that the TTO imposes unreasonable standards or is very slow in acting.
7. A discussion with the academic administrator to which the investigator reports to determine this person's general thoughts on entrepreneurial activities and whether this person is familiar with any similar activity in the unit, college, or other immediate academic areas.
8. Discuss the climate of the unit as regards peer opinions of entrepreneurial activities. The potential practitioner may find a sizable percentage of faculty who view any activities in which the faculty member or any company involved in the technology licensing as objectionable and not worthy of the particular institution. Others may view such activities as marginal. These attitudes will be important to a potential entrepreneur since if negative attitudes are prevalent, there is unlikely to be peer support within the community of academic colleagues. In that case, pursuit of entrepreneurial activities still is possible, but the person involved will be isolated in some degree. If this is the case, then the person involved has to decide how much they to pursue these goals. Isolation is uncomfortable but it still is still possible to proceed, as the author's experience demonstrates.
9. Examine the stage of career of the potential entrepreneur. Unless the position description specifically demands such activities or the reward structure of the university is supportive as described above, then nontenured faculty are at significant risk. The evaluation committees will be focused on the specific position description and evaluate according to the reward structure of the university. Therefore, a nontenured person can easily put themselves at risk of their academic career if they pursue entrepreneurial goals at the expense of other written job duties.

10. Evaluate the time commitments required for both standard academic activities and entrepreneurial activities. For example, if a person has a heavy course load, then there probably would not be time for entrepreneurial activities. Of course, any investigator or faculty member with such circumstances may not have generated the ideas and potential inventions that would support an entrepreneurial role. The potential entrepreneur needs to consider the role he or she would play in any organization with which he or she is affiliated. As described in Chap. 4, it is almost certainly an intolerable conflict of interest if a full-time academic wished to be the operating officer of the president of any for-profit company. There are potential roles for an academic in nonprofit and/or grant-driven organizations.
11. The entrepreneur needs to consider whether his or her entrepreneurial drive and the opportunity presented is sufficient for him or her to be willing to either resign from the institution or to become a part-time employee. Clearly, if the entrepreneur resigns from his or her academic appointment, then there are no conflict-of-interest issues to consider. If proper disclosure is made, then there may be an opportunity to pursue activities such as those just described on a part-time basis. A split between academic and commercial pursuits may be permanent or temporary, but this must be determined by discussions with appropriate academic officials before any changes are made. A temporary part-time split may or may not be acceptable to the institution. The author of this book, as indicated in the next chapter, reduced his tinge to about 55% at Cornell and spent the remainder of his time as the CEO and/or Chief Scientist of a company that he founded. After about 2 years, he returned to his full-time employment at Cornell. Similarly, at MIT, faculty frequently reduce their academic appointments to part-time to pursue for-profit activities related to their academic research (Shane, 2004). This may or not be acceptable in any given academic institution and it may depend upon the duties and requirements of the institution for the individual involved. Flexibility on the part of the institution to permit such activities may be a powerful inducement and requirement to entrepreneurship on the part of its faculty.
12. What are the opportunities and pitfalls to fund translational research and product development?
13. Are there institutional funding opportunities specifically directed toward activities that create jobs or wealth or that are specifically designed to translate basic science to useful projects. As described in Chap. 2, there frequently is a large chasm between basic scientific discovery and product sales. In many institutions, there are funding programs directed toward crossing the chasms. At Cornell, and elsewhere in NY state, the NYSTAR program funds centers for various purposes. These state funds require a match in cash from a NYS company but has as its mission creation of jobs for the state. The author has received many of these, and while the amount of funding is rather small for each grant, the total in fairly continuous funding was very critical. At the University of Rochester, there are grant programs specifically to bring products to market and these are supported by gift-giving (endowments) for this purpose. Thus, the

University of Rochester is willing to use endowment giving to support entrepreneurial endeavors. Cornell, at least in the recent past, was very reluctant for the TTO to provide names of donors as potential funding investors for a fledgling start-up in the fear that this would undercut capital endowment building programs.

14. Are there local investment groups that may be interested in the new company? Are these university-based, and, if so, is the new idea something that these local investment groups would be willing to support? Some universities, including Cornell and the University of Rochester, have small investment groups largely run by students with faculty advisors that make small seed investments in selected companies. There may also be business plan competitions with significant (at least at the early start-up stage) prizes to the winning groups as well as regional groups that support wider scale competitions and opportunities for investment approaches.
15. Does the institution permit grants and contracts from companies in which the principal investigator has a financial interest? This was discussed in the last chapter and is an essential need for many circumstances, but this is not universal. MIT, for example, does not permit grants or contracts in those circumstances. However, they have a very strong infrastructure of investors in the Boston area that are very accustomed to funding MIT-based start-ups. As a consequence, investigator-based start-ups are common, but typically the start-ups are funded and rent labs outside of the university and the involved faculty members take a full-time or part-time leave (Shane, 2004). In many other locations, such investor-financed labs and salaries are not possible because of the lack of financial resources in the local area. Thus, specific models will differ from place to place and institution to institution.

What are the institutional fringe benefit, overhead, and other rates? Assuming that grants or funding from companies are allowed, these costs become very important. At the end of the author's career at Cornell, the fringe benefit rates were approximately 50% and the overhead rates (indirect) were another 50%. These rates are calculated, and the subject of negotiation with organizations such as the National Science Foundation, but most companies, especially in the agricultural arena, especially small ones, will find such costs unreasonable and so will be very reluctant to do product development in the university laboratories that include high-cost structures. Some methods whereby such impediments can be overcome need to be implemented. One possible way, as will be discussed in Chap. 5, is grants. There are a number of foundation and were approximately 50% and the overhead (indirect) costs were another 50% of the total. Thus, to hire a person with a direct salary of $35,000, the total that must be raised totals nearly $70,000. These rates are calculated and subject for negotiations with organizations such as institutional grants which are available to companies, and they permit a certain portion of the R&D to be done in universities as subcontracts. These subcontracts may mandate a lower level of at least indirect costs than the university's full rate. Is it possible to negotiate a lower rate, but in the author's experience, this was impossible.

There has been a novel and useful structure at Cornell that helped to deal with a high-cost structure. The Cornell Center for Advanced Technology (CAT) Program, which also exists at other established centers of excellence in NYS, provides grant funding. However, there must be a NYS company that provides matching funds. This matching funding augments the company dollars and so, even though high overhead is charged on the company matching funds, the funding from CAT at least partially offsets the high overhead and makes progress and funding possible.

What additional support can the university extend to the potential entrepreneur? As implied in the two quotes at the beginning of this chapter, different institutions have very different policies. It depends on whether the institution "gets it" and whether or not it is supportive of such activities.

> One important aspect was suggested at the outset of this chapter: if a university investigator is thinking of starting an entrepreneurial project, whether for commercial gain or not, it is helpful if there are advisors who may advise the person on promises and pitfalls of the proposed venture. These frequently are experienced business persons with good connections and in most places have the title of entrepreneur-in-residence. Many institutions, including Cornell, have such persons and it is very useful to a fledgling potential entrepreneur to discuss the potential program with them. These individuals may be able to advise on creating teams (Chap. 5), funding the project or company, and legal affairs (Chap. 6). All of these are critical items for the establishment of an entrepreneurial organization or enterprise.

1. Are there interns, students, or others that can be employed at a relatively nominal cost as part of their academic training? Investigation of the availability of such persons to perform specific tasks can be highly cost-effective. Many institutions have such programs that can be available. Work-study programs for student may be another possibility.
2. Are there training grants offered for new businesses by local or state authorities? In NYS and the county where the author resides, there are several programs that offer assistance for hiring of persons who meet various economic or other criteria. These may provide up 50% of the salary of the person hired for a period of a few months while the new person is being hired. This can defray total costs when even small outlays are critical for a new program.
3. Are the local, state, or other programs willing to assist with costs and to get programs running? Many or most governmental units offer incentives of various sorts, ranging from outright grants to specific job-based incentives. Some examples include:
4. Tax breaks for start-up companies: NYS offers tax-based incentives for job creation, for example. These incentives provide for a set amount of value for each job created over a certain minimum.
5. Incentives for new business and energy efficiency improvements for new construction. Some offered by the local electric and gas company provide for a substantial portion of the cost of new electrical infrastructure and other programs for installing energy-efficient lights and appliances.

6. Training grants for new employees. The local county provides training grants for companies or organizations that employ low-wage persons that can be trained to do more demanding tasks. These frequently may provide a significant portion of at least the first year's salaries.
7. Are there entrepreneurial courses offered by experienced faculty in the business or related schools? These may provide invaluable information that may augment that provided by persons such as entrepreneurs-in-residence. The students themselves may also be candidates for the teams that are an essential component of any entrepreneurial program (see Chap. 6).
8. 14. Finally, discuss any start-up in detail with the TTO. This office may be able to help with many of the points just discussed. Of course, a conversation, as mentioned earlier, regarding the university patent policy is essential if patenting is part of the strategy being employed.

Universities Must Keep Faith with Their Entrepreneurs

In at least some cases, royalty streams are essential if the program is going to have sufficient funding to operate. Many programs have insufficient opportunity to generate large sums of revenues from grants. These same programs, which are providing useful and economic values to society and university, frequently cannot pay high overhead and fringe benefit rates such as those described above.

Typically, at least at Cornell, royalty payments are paid approximately 1/3 to the patent office, 1/3 to the college and university, and 1/3 to the inventor. Of the amount paid to the college, approximately half came directly to the department or the program generating the income.

These funds typically have been free of the overhead rates on standard grants. This frequently is about the only source of funding for these programs (see Chap. 5).

A few years ago, however, the following occurred:

1. The college took much of the funding that had been flowing to the program, leaving only 10% to the program that generated the revenues.
2. The college justified this by stating that the inventor could donate their personal share to fund the program, and many did.
3. However, later the university confiscated some of these funds, including those that had been directly paid from inventor shares.
4. This clearly had a chilling effect upon inventors providing their share of royalties to fund their programs.

Therefore, these programs had a very difficult time surviving. Moreover, this has the very likely effect of cutting off the revenue stream from these programs to Cornell, since programs cannot survive. This is a clear case of "killing the goose that lays the golden eggs" and is exactly what ought NOT be done by any entrepreneurial institution that has any hope of increasing royalty and similar revenues. No wonder some faculty suggest that "no good deed goes unpunished."

The following is a proposal that the author believes would go a long way toward making the institution friendlier to the entrepreneurs and that would benefit the institution:

1. The university encourages or at least offers the possibility of the inventors of inventions giving their share of royalties to support their programs.
2. This would be much more palatable to inventors, and better for the institution, if the university and/or TTO matched the inventor royalty contribution with funds from the institution.
3. As was just noted, very little of the total royalty revenue goes to the program. If much more of this total was provided, then the program would prosper, more royalty-bearing IP would be created, and both the program and the university would benefit. Instead of killing the goose, the goose would be fed, and the eggs would be larger, more frequent, and more golden. Everyone would benefit.
4. It is the author's belief that this cost-sharing model would be popular and that the scientists involved would have a much greater sense of fairness and receptivity of the institution to their research goals and aspirations.

In terms of comparison, again using the small fruit-breeding model, the University of Florida provides more than 50% of the royalties generated back to its program. This compares with 10% at Cornell. The program at Florida is prospering and generating increasing levels of royalty dollars back to its small fruit-breeding program, while the one at Cornell is in difficult circumstances due to the lack of funding. It is within the power of Cornell and its TTO to rectify this situation easily and to increase the fairness between the investigators and the institution. Certainly, the author would have very much appreciated such an arrangement. This would also benefit the shareholders, such as homeowners and businesses that rely on university-based improvements.

In short, it is important to discover the resources that are available to any new entrepreneur. As both governments and institutions become more entrepreneurial, they are beginning to vie for services and advantages that are useful to create a more competitive economic environment. Interestingly, in some institutions there are both positive and negative forces that countervail each other, and in investigating the advisability of going forward, it is prudent to examine as many of these as possible.

Lessons from Chap. 1

1. Many academic institutions are retrenching because of financial difficulties. There are numerous reasons for this including decreased public funding, increases in unfunded mandates, and increasing levels of administrative staff, partly to meet the unfunded mandates.
2. As a consequence, when faculty members retire, many of the positions are not filled. Additionally, to increase tuition revenues, the class hours frequently

increase. These two, plus others, are degrading the abilities of universities to maintain highly robust and innovative research programs.

3. Academic institutions are attempting the difficult job of maintaining quality institutions. However, in some cases, there is an attempt to optimize existing resources, when innovation into entirely new directions may be a better path. The effort to optimize resources frequently includes dramatic increases in academic unit size. Usually this includes a cadre of secondary administrators below the department head or chairman, increases the level of bureaucracy, and decreases the commonality among peers in the unit. In the author's opinion, this leads to greater conformity and has a stultifying effect on faculty morale and creativity.
4. The author believes that universities and other academic institutions are unique and need to embrace the following principles:
   a. Universities are among the primary developers of intellectual capital in most of at least the developed world.
   b. A great deal of the discoveries and inventions at universities can be crafted into useful products or services that benefit society, both in terms of the betterment of society and the important financial advantages provided.
   c. Universities should have and accept the obligation to create jobs and foster economic growth for the taxpayers and others who have supported their activities.
   d. For reasons already described, many universities do a poor job of fostering useful entrepreneurship.

References

AAUP. (2012). *AAUP recommended principles & practices to guide academy-industry relationships*. pp. 307.

Shane, S. (2004). *Academic entrepreneurship: University spinoffs and wealth creation*. Edward Elgar.

Chapter 2
Reasons Why a University Faculty Member Might Want to Become Entrepreneurial: Whether With or Without Involvement of a Commercial Entity

Why Universities Need Entrepreneurial Faculty

The science developed at colleges and universities is a driver of economic opportunity in the USA and around the world whose capabilities can be increased. Globalization of the worldwide economy "means that the only way that developed western economies can compete with lower paying economies worldwide is with technology, talent and entrepreneurship" (Allen & O'Shea, 2007). Three requirements for technological economic development are the presence of high-quality universities and research-oriented engineering schools, an inviting living environment, and the availability of venture capital (Goldman, 1984).

Given this statement, it seems obvious that universities and other publicly or privately funded research institutions have a responsibility to ensure that the technologies developed by their faculty is (a) made available to the public and (b) generate economic activity that can help fuel the economic engines of the nations in which they reside and to provide revenue for the university. Unfortunately, a large, and undefined, proportion of university research ends as scientific publications but is never translated to useful and usable products or technologies. Therefore, much of this does not generate economic value to their societies or to their universities other than indirect costs they generate on the research grants they obtain.

At the same time, academic institutions in the USA, and probably around the world, are in crisis. The public institutions are caught in a downward spiral of lower public funding and increasing costs. This change occurred earlier in other countries. Private institutions have the same issues, but because they are generally more expensive, they may be better able to weather financial crises. Both private and public institutions are responding by increasing costs to students—in private universities, tuition costs have doubled between 2000 and 2018. Since 1990, faculty numbers in many publicly funded universities are declining because of cost pressures. This has occurred even though student numbers, and therefore teaching loads, increase. This

G. E. Harman, *Academic Entrepreneurship*,
https://doi.org/10.1007/978-3-031-06821-8_2

means that a smaller percentage of faculty time is spent creating and innovating (i.e., doing research). These changes are occurring even though research expenditures to US universities, hospitals, and research institutions were at $53.9 billion in 2009 from $27.9 billion in 2000 (Anonymous, 2009). All of these changes and more must lead to a decline in research productivity in the USA and elsewhere. This is in direct contrast with the statements in the first paragraph that states that universities need to provide the technical leadership that leads to economic development.

All of this suggests that universities may need to seek new or alternative ways both to enhance the fulfillment of their implied contract with the society that supports them and to generate sufficient revenues to maintain themselves as strong and vigorous institutions. Typically, at least for US land grant institutions, the universities are a three-legged stool—teaching, research, and extension of knowledge to society. This book challenges universities and research institutions to embrace a fourth dimension—entrepreneurship. The universities with which I am familiar provide salaries to faculty (a few do not even provide this) and a roof and a lab, and everything else for at least their research program has to be generated by the faculty, including all personnel salaries (usually), equipment, supplies, and service costs. This is typically accompanied by high fringe benefit and overhead costs. This requires, in fact, that faculty, in some fashion or another, have to be highly entrepreneurial.

As mentioned earlier, Drucker (1993) defines entrepreneurs as ones that "create something different; they change or transmute values." He also points out that "nothing could be as risky as optimizing resources ...where the proper course is innovation." Universities today are doing their best to optimize their resources, which includes simply consolidating and moving units and resources to accomplish the same goals as before. To obtain their operating funds, the preferred options are increasing student tuition, overhead on grants, and increasing endowments, largely from alumni. Moreover, to an increasing degree, universities are becoming more regimented, are increasing administrative support staff, and increasing unit sizes. This increasing level of regimentation and increased unit size runs counter to what the author views as critical—creating greater entrepreneurial opportunities and grassroots efforts from faculty. Across at least the USA, the view of many or most faculty is that the environment for creativity is becoming more limited and that the general atmosphere is stultifying. Smaller groups with similar foci generally are more creative. One of Cornell's greatest strengths used to be "creative chaos," which was an atmosphere in which unusual and creative ideas could easily emerge. This strength seems to be waning and increasing bureaucracy taking its place.

A potential method to innovate, in Drucker (1993), is to embrace entrepreneurship. In terms of this book, this equates primarily to creating an environment where, as a primary motivating factor, technological advances are transformed into useful products and services that create wealth in the world, the nation, and in the communities in which they reside. The methods to increase this wealth and well-being needs to be fostered and encouraged. In many, but not all cases, this requires improved methods and especially university cultures that encourage entrepreneurship, and where appropriate, patenting and, then commercialization of the patents that they generate. If they do this, then the following will occur.

- The universities will generate significantly more royalty revenue from inventions by faculty.
- They will create jobs and wealth around the world. Colleges and universities do create jobs *and* wealth, but typically they do not quantify jobs or wealth nor do they provide rewards for faculty that are effective in this regard.
- They will increase funding opportunities for their faculty, which is a critical measure of success. The old "publish or perish" paradigm has been replaced by "get funding (preferably with high overhead rates) or perish" because there are no other ways to generate successful research programs.
- They will provide opportunities for faculty to improve their financial circumstances by royalty sharing on inventions and by providing direct opportunities for financial gain.

The first item above needs amplification. Universities and other research institutions have significant OTTs (offices of technology transfer). The OTTs provide valuable services and are responsible for administering technology transfer including managing patents and licenses. These generate significant revenues for universities and, by extension, create significant increases in employment and wealth. In the USA, the Association of University Technology Managers (AUTM) records and chronicles activities. In 2009, the total license income to all universities was $2.1 billion. The majority of this income was from licenses on patents so the income to universities was large. This income is not equally distributed; of the 179 universities that responded, the top 25 accounted for 83% of the total income. This certainly must translate into a substantial amount of wealth creation and jobs, but that data on job and wealth creation is usually not collected by most universities. This is surprising since legislators frequently seek information on this topic to defend university budgets for public institutions. Specific data on job and wealth creation is available and should be reported. Necessary data is provided by university researchers. If very specific job creation numbers were available to university administrators and could be conveyed to legislators, this would go far in defending public research budgets. This would be especially true if universities created a climate fostering entrepreneurship, job, and wealth creation, which is not currently the case.

Moreover, if we assume that, at a maximum, the royalty income to universities accounts for 3% of the total economic returns to companies licensing the technologies, then the total economic benefit from the licenses is $70 billion annually; in the year 2000, the benefit was estimated at $33.5 billion (Cohen, 2000). This value derived from AUTM probably is an underestimate because this organization only records and uses data on intellectual property that has an initial license fee of $1000 or more and many licenses do not meet this requirement, even though they may become highly successful (communication from Alan Paau, Director, Cornell Center for Technology, Enterprise, and Commercialization). In the author's area of upstate NY, just two institutions (Cornell University, including the medical campus in NYC and the University of Rochester) received about $70 million in royalty and related income according to their most recent annual reports, which using the same 3% rate values would result in $2.3 billion in economic activity. This is a significant return, but one that the universities with which I am familiar do not publicize nor use to their advantage. It is the author's contention that this return could be greatly augmented by a change in policy that encourages and rewards faculty inventors who create new economic opportunities for taxpayers, students, granting agencies, and others who support them. University start-ups also have a high success (survival) rate. The survival rate of academic start-ups is around 70%, while for start-ups in general the 10-year survival rate is only about 30% (Shane, 2004, 2010).

Moreover, it is the author's strong belief that at least many university cultures and administrative structures are unfriendly, or at best, indifferent, to faculty that generate patents and that create royalties to the university. A more proactive policy toward royalty generation, with appropriate faculty rewards, could markedly enhance royalty income. One of the author's patents, over its lifetime, generated about $2 million in royalties and this was at that time the highest grossing patent in Cornell's College of Life Sciences (CALS). Given the capabilities of the CALS faculty, this is a tiny amount of revenue over the 15 years or so of royalty generation. A more proactive structure, with an appropriate reward structure in terms of advancement and other credits to the academic, could greatly increase this level of return to the university.

Of course, the information just presented is based on patent and related revenues and does not include economic advantages created by non-patented entrepreneurial activities. In Chap. 2, there are three examples of specific programs that do not involved patented activities, at least not primarily, and these also generate significant economic benefits to the community and to the world, even though funds from royalties do not occur. However, universities need to collect data on economic value and job creation, from entrepreneurial activities. The universities need to be able to tell legislators and other funding sources, including donors, what economic advantages that the university creates. This can be reporting done by the investigators and verified by the university.

Entrepreneurship, as defined in this book, can assist universities in meeting the societal needs as described above, it can assist in funding if programs are put in place to encourage patenting and royalties derived therefrom, and it can assist faculty and others in meeting the financial requirements of their research programs. However, in many cases, this requires changes in university policies and the institutional environment to assist, rather than hinder, entrepreneurial faculty. Some institutions have a high percentage of faculty with entrepreneurial programs, while others have very few. This largely is a consequence of administrative policies and university culture, and this book seeks to define the policies and practices (see Chaps. 2 and 5) that encourage or discourage entrepreneurship.

The Difficulties of Commercializing University Technologies

There are different types of challenges in commercializing university technologies. In some cases, highly innovative entrepreneurial programs can be conducted and financed without a company or other entity directly involved in the program. In these cases, commercialization and value to society can proceed via established market channels through partnering with public and private institutions or organizations. This was the process by which most of the first green revolution occurred, and it is still viable today, albeit with vastly improved tools for discovery and development. In other cases, university or academic research produces inventions or discoveries that cannot reach the public because they represent totally new product categories. In this case, the inventor and his colleagues needed to develop commercially acceptable products, manufacture them in sufficient quantities for the marketplaces, and finally market these. The discussion that follows categorizes these types of technologies as adaptive or disruptive, and the case studies in Chap. 3 give examples of these. Both are quite important; in some cases, the only way that university technology can be implemented and provided to the world is through commercially based entrepreneurship, while in other cases, more classical totally university-based systems are appropriate.

There are fundamental differences in the abilities of many universities and other similar institutions to deal with these two categories. Adaptive technologies fit easily into the standard university modes of action. However, systems for disruptive technologies, which typically involve a mix of university and commercial-based practices, are a more uneasy fit into universities. Conflict of interest management and even, in some cases, peer pressures are likely to be much more of a burden.

In the previous chapter, I discussed adaptive vs. disruptive or transformational technologies and the difference between them.

1. *Adaptive technologies*: These are systems that enhance an existing product in an innovative and superior way. In this case, the basic product being sold is still similar to existing ones and a market already exists. An example could be a new and innovative camera lens or light metering device. In the examples in Chap. 3, the examples of improved rice yields and systems and the research to allow broccoli to be produced in the NE USA are examples of adaptive research. The products being sold are still rice or broccoli or a camera. It may indeed be possible, as in the examples, to have a strongly entrepreneurial program but without, in the words of McCouch, the role is not about generating revenue. This is possible because there already is an established marketplace for rice, and if there is more and better quality rice, the marketplace is ready to absorb this, or the farmers who directly consume it will do so with an accompanying improvement in diets and societies. Such activities that are funded totally by public or foundation sources and that work directly with farmers, producers, and others in the agricultural sector are typically in the mainstream of a university, especially one that deals in agriculture.

2. *Disruptive or transformational technologies.* This category deals with products that are new to the marketplace and, in many cases, would not be recognized as marketable or a public good until they were driven into the marketplace and became part of accepted society. There are many examples of this sort—to use the camera example, the change from film-based imaging to digital systems is a case in point. Parenthetically, this example also demonstrates the difference, in Drucker's terminology, between optimizing resources and innovation. The Eastman Kodak Company did its best to optimize film-based resources when they could have become an innovative leader in digital imaging. They missed that opportunity and the results are very obvious, a major company that almost died. Other examples include cell phones and computers—when the early pioneers in small personal computers began working, the large companies all were focusing on large central units. This created an opportunity for the garage-based entrepreneurs like Hewitt, Packard, Jobs, and Wozniak. In Chap. 3, there are two examples of such disruptive technologies—the first is the development of the "gene gun" which was a device to insert foreign DNA into various organisms by actually firing a projectile into tissues of the organism, and the second is the development of endophytic microorganisms that colonize plant roots and induce changes in plant gene expression that result in improved plant performance and yields. All of these very diverse technologies absolutely require a commercial entity since in every case, a new product concept had (a) to be invented, (b) produced in commercially acceptable form, and (c) the products had to be introduced and sold into the marketplace. This process absolutely requires a commercial entity, and the company has to generate operating funds and there must be a profit motive, or no one will be willing to invest the funds required. Funding also requires that the technology or product be protected, usually by patent filings. In every case, these disruptive technologies were unexpected and involved totally new products and ways of doing things. If the authorities in established companies were asked what was needed in their field, in almost every case, they would have opted for an adaptive, rather than a disruptive technology. In most of these cases, the new products were the vision of only a few, or sometimes a single individual, and were never the product of committee or a large group of established experts. Such technologies are more difficult to develop to fruition, and the reasons for this will be discussed later in this chapter. They also fit much less comfortably into academic settings. By definition, they require a profit motive to bring to fruition, and this is undesirable on the part of some academics. The inventor has to drive the technology (see Chap. 3) and he/she must also have "skin in the game" in terms of his or her own investment and must also expect to make money. If the inventor does not have a dedication to the profit motive and see the opportunities for doing so, he/she will never convince investors to become involved. Thus, inventing and driving a disruptive technology are likely in general to more difficult, since skeptics must be convinced of the worth of the new thing. In addition, since the profit motive will be a necessary component, conflict-of-interest issues must be considered, and all regulations must be met.

Motivation of Academic Scientists for Entrepreneurship

In this chapter, we will deal with the reasons why an academic scientist would want to develop an entrepreneurial program, whether adaptive or disruptive. We will examine where difficulties lie; the intrinsic reasons why there is difficulty from the point of view of the scientist, the university, and any commercial entity; the economic advantages to the university scientist in doing so; and whether or not an academic scientist will personally make any money, if this is a factor in deciding how to proceed.

Interestingly, in large part, entrepreneurial faculty-based (or even outside academe) programs, personal gain is not a primary reason for choosing this route. As noted especially in Chap. 5, there are types of entrepreneurship that do not involve company or the profit motive, but in large part, the issues described in this chapter pertain to them as well.

Even scientists who have major financial success, this is the case. Ray Ozzie, who was a University of Illinois faculty member, developed an early collaborative software application. Ozzie later formed Iris Associates to develop Lotus Notes for Lotus (Livingston, 2007). Lotus itself was purchased by IBM for approximately US$3.5 billion. Approximately US$3 billion of this price is generally attributed to the purchase of the Notes business (Livingston, 2007). Later, Groove Networks, a software company founded by Ozzie Groove, was purchased by Microsoft for an undisclosed sum. Microsoft marketed the collaboration software as Microsoft SharePoint Workspace (http://en.wikipedia.org/wiki/GrooveNetworkq).

Even so, he stated that "Everyone knows that one reason you go to work is the hope that eventually you'll be compensated. But you don't have to say it....It should be about the mission. It should be about changing the world. It should be about how you can impact the lives of users, partners" (Livingston, 2007). This desire to make the world a better place is an important aspect of many or most entrepreneurs.

Professor Susan McCouch of Cornell University has a large and important program to feed a hungry world. Her attitude is expressed in the following quote: "We work hard to partner with those who have a responsibility to feed hungry people." "The role of rice research at a university like this is not about generating revenue—it's about the public mission of Cornell." Her program is highly entrepreneurial and described in more detail in Chap. 3. However, it has nothing to do with company start-ups or personal gain. In the two examples just noted, Ozzie made a considerable amount of money directly from his entrepreneurial activities, and McCouch will seek no remuneration directly from these activities (other than her faculty salary and other perks). However, both have a strong motivation to improve the world by the application of their talents and the technologies they develop. The routes and paths they chose, along with the teams of like-minded persons who they recruited for their teams and colleagues, are quite different in both approach and motivation, but the same type of mission and desire to improve the world were the primary components of their efforts.

Livingston (2007) describes attributes of entrepreneurs; the characteristics he describes are similar, regardless of whether persons involved are in academe or not and these seem to the author to be correct. He says that some kind of magic happens when startups or other entrepreneurial programs first begin, but the only persons who know about this are the founders. One attribute that surprised him is how unsure founders were that they were on to something big, which is another way of saying that it is impossible for the founders at the beginning to have any assurance of success or even survival on the entity that they are starting. However, a distinguishing characteristic is that they were all determined to build things that worked and that this determination is the most important quality in an entrepreneur. The determination of the entrepreneur to change the world, as indicated by Ozzie, is a critical factor in success of start-ups. However, there has to be a high tolerance of uncertainty. Steve Wozniak, who was the primary engineer in the early development of Apple computers, said it well: "Entrepreneurs have to keep adjusting...everything's changing, everything's dynamic, and you get this idea, and you get another idea and this doesn't work out and you have to replace with something else. The reason I had it so intense in my head, and the reason for that was largely because is was part of me. The computer was me" (Livingston, 2007). Clearly this path can work; Apple computers are highly successful but a high degree of confidence in the entrepreneur's problem-solving ability and a tolerance for uncertainty is important. Similarly, McCouch's program in rice is highly successful now, but it is very unlikely that this was clear from the beginning of her work. The ability to persevere in the presence of uncertainty and to have the zeal and energy to overcome barrier after barrier is the hallmark of any successful entrepreneur.

Of course, this uncertainty is a significant factor not only to the faculty entrepreneur but also to the institution for which he/she works and for investors and colleagues who decide to join the team to create the entrepreneurial vision of the specific product or service being developed.

For academics, an important driver may be the wish to create something of lasting value. The standards for success in academe are teaching, publishing, and obtaining funding and, for some, extension, i.e., demonstrating techniques to users of technologies, and not necessarily their own. Certainly, there are rewards in the accomplishment of these goals. One of them is a successful cadre of students that the faculty has mentored, awards from peers, and the like.

However, if one does research, it is important for many of us to actually see what we developed used in a way that makes the world a better place. In many cases, this does not happen, even if there is a legacy of a large body of published work that only a few read and that, as technology progresses, is only a footnote in the long history of science.

For the author of this book, and most other entrepreneurs he knows or has read about, it became personally important for the technology being developed to actually be used by real people in real-world settings. Papers in a journal, no matter how prestigious, were inadequate. The area of research (see Chap. 3 for details) was to develop microorganisms that help plants ward off diseases, improve productivity for a hungry world, and similar concepts. Someone has actually to go and put the technology into practice if it is going to be of benefit, either for societal good or to create monetary value, including jobs, royalties back to the university, or local economic development. It has been quite a nice affirmation to see the microbes that the author developed go from a laboratory curiosity to large-scale production and to take a solid and increasing role in worldwide agricultural practice because they provide new and lasting value.

A logical question then becomes: once the technology is developed at the university, why not simply license the technology to an existing company? Why does an entrepreneur need to go to all the bother and difficulty to get the technology deployed and used? The short answers are that few or no companies will take on a new and unproven technology and the inventors the one most suited to carry the technologies forward. The reasons for this are provided in the next section of this chapter.

As an example, the first technology commercialized by the author was a microorganism developed for biological control of plant diseases. The first attempt at commercialization was to license to a large company, in this case, Eastman Kodak in the late 1980s. Kodak expected to use this strain as their initial entry into agricultural biotechnology; this was part of a larger attempt at diversification which include potential entries into drugs and agricultural chemicals. They spent a significant amount is this attempt, and then with financial reverses (e.g., the settlement of the Polaroid patent infringement case) they abandoned all of these forays. With T22, they obtained an important regulatory approval and then turned the entire package back to Cornell University's patent office. As described more fully in Chap. 3, this entire package was licensed by the author and two cofounders of what became BioWorks Inc., which is a profitable company still located near Cornell University.

In hindsight, the failure of the Kodak venture was predictable and is instructive in this book. First, large companies frequently are not entrepreneurial—Kodak had a tremendous investment in film technologies and wanted to optimize these investments rather than evolving into digital systems where they already owned the patents. Kodak could have become the major leader in digital photography; many of the first patents in this field were owned by Kodak. This is not dissimilar to many other large companies with substantial legacy labor costs and with a strong investment in their existing technology. As long ago as 1993, Drucker (1993) predicted that "today's businesses, especially the large ones, simply will not survive in this period of rapid change and innovation unless they acquire entrepreneurial competence." Entrepreneurial competence was well defined in his book but is probably seldom practiced by companies large or small. Then when they did try to innovate as in the case of T22 or their investment into Bayer aspirin and related areas, they tried to do so in a field where they lacked expertise and market presence. Kodak management had few skills in the agricultural arena and hired marginally qualified, and certainly not entrepreneurial, management persons. They also, in that time frame, never fired anyone, and when a new division was organized, they promoted persons who did not fit into other organizations within the company. In short, Kodak management that staffed and organized their fledgling agbiotech division had little or no experience it the field and lacked an entrepreneurial management plan. This is of course ironic because George Eastman was a great innovator, with his development of flexible film systems replaced cumbersome glass plates and so made photography accessible to the public. Similarly, much of the early development of computers was made by enthusiasts in clubs that worked for the love of the challenge. So, in this case, success was unlikely. It also is fair to say that there was essentially no capabilities or knowledge of how to scale up, test, or market T22 either at Cornell or at Kodak. In fact, at that time, those skills were not available anywhere in the world.

Thus, even if licensing is possible for a new technology to a company or other party with the financial resources to bring a technology to market, these large companies may be poorly suited from internal structural reasons to make the technology successful. One reason is that there may be a large time lag between first discovery and eventual use and commercialization.

Another difference may be as fundamental as why the technology was developed in the first place. Businessmen perceive value in R&D as producing a product that the market wants, i.e., is an adaptive technology that will enhance an already-profitable product line.

Computers and digital cameras are examples of visionary concepts that were world-changing but, at the time they were developed, it would have been difficult to understand how they fit into immediate market wants. Such technologies and products are the province of visionary entrepreneurs and are not in the scope of big business. There ought to be no one better place to provide visionary products, whether adaptive or disruptive, than visionary university scientists. In this climate imagination and vision are staples and the creative process can flow well if the institutions allow and foster such development.

The Valleys of Death for University Technologies

Universities, and their funding sources, are excellent at developing basic science. In the agricultural fields, especially, there was a very strong extension service that delivered technological insights and that provided an honest appraisal of different technologies, varieties of crops, and the like. This extension service, which was funded by a state-federal partnership together with local governments, had a major role in the creation of the USA as the breadbasket of the world. This capability is greatly eroded at the present time and is largely funded by the states and local governments.

However, this solid basic science capability of the USA and the rest of the world frequently does not result in products that are usable. Figure 2.1 shows in diagrammatic form some of the issues and why this is so. At the stage of technology development that typically results in publications and will support a patent, the technology is at the descriptive, fundamental stage. Much is known about the basic science and fundamental basis of the discovery.

Unfortunately, at this stage, the technology/discovery is very far from being a usable product. Typically, little is known regarding how to formulate the discovery

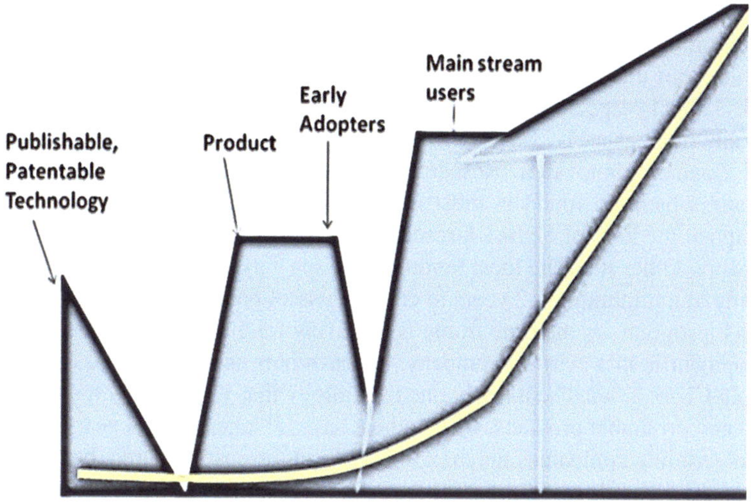

Fig. 2.1 The valleys of death that occur on the way to commercialization. Publishable technologies are usually not ready for commercialization—once the original concept has been developed, methods to produce, market, and manufacture must be done. The first valley of death is the transformation of concept to commercial reality. Even after this is done, buyers must be found. The first to purchase are the early adopters, but they are unlikely to be repeat customers, instead they are likely to move to the next new thing. Funds to bridge these two valleys must be found. The yellow line indicates levels of revenues likely to be available. Chapter 6 discusses possible sources for the necessary funds. This figure was based on Moore (1991) (1). A similar illustration was published in Harman (2019)

into a usable product or a product system, how to produce the technology on a large scale, and how and if it fits into the marketplace. For these reason, even excellent technologies require considerable amount of time before a useful product and methods to use it are developed. In many cases, extensive trials and evaluations have to be done and, in some cases, such as pesticides, some pollution control agents and medicines and medical devices may require a rigorous and lengthy registration process. These reasons all contribute the first valley of death, which is the period between discovery, publication, and patenting and the actual creation of a useful product (Fig. 2.1). This is especially true of disruptive technologies but it applies as well to adaptive systems and discoveries. McCouch states that it took 20 years and 34,000 attempts to cross a domesticated rice variety with a distantly-related salt-tolerant wild relative from India before fertilized offspring were obtained. It will now take at least 4-5 years of breeding to eliminate wild characters to generate a new high-yielding salt-tolerant rice variety (McCouch, 2013).

In the rice example just given, Professor McCouch has enjoyed strong support from granting agencies. In other cases, there is very little public funding for development of especially the most innovative and/or disruptive systems since usually there is substantial skepticism whether the product or system has value. In 2001, the grower magazine Corn Farmer discussed one of the author's first biological agents with this title "Is T-22 Foo Foo Dust or the Real Thing?" Similar skepticism was the norm at granting agencies as well. It is now clear that the products are indeed the real thing, but this doubt does not help funding, either from commercial or public sources. This is especially true if the technology flies in the face of accepted wisdoms and dogmas as T-22 and related technologies did. However, in the USA, the Small Business Innovation Research and Small Business Technology Transfer Program, which are found in most federal granting agencies, provide funding to accomplish the kind of high-risk research that is required for this type of development work. Other state and local funding agencies may also be of assistance. It takes typically at a minimum 1–2 years to create a useful product from a basic academic research program, so the time frame is relatively long and, during that time, if the development is in a start-up company, the cash burn is ongoing and is difficult to meet. In Chap. 4, we discuss only one technology that very rapidly transferred into useful and profitable products, while others have all taken several years. Grant support to fledgling companies may not only help in the cash required, but it may also validate the technology for later investors. This is one reason why, as discussed in Chap. 3, university funding based on commercial successes is an important component of the necessary infrastructure.

Clearly, fund raising during initial product development is not an easy process. However, in some cases, it has proven essential to finding the funds to produce a product that would not have been possible through conventional university grant funding. Shane provides several examples of this, and the author could never have developed the microbial products that have proven effective without a mix of both university- and company-based financing.

Figure 2.1 also shows a second valley of death. Once an original prototype product is produced and introduced into the marketplace, then there must be a market created for the product. Products typically are first purchased by early adopters, which are the risk-takers of the customer base. Unfortunately, even those early adopters are also likely to move on the next flashy new product and therefore not to become repeat buyers. It takes hard work to move from the early adopters to the mainstream users, at which time the product becomes secure in the marketplace. Moreover, it is also almost certain that the initial product will need to be improved. Few first market entries are perfect (look at Boeing's 787 aircraft), and there probably will be improvements that meet the needs of the marketplace than the first introduction. Creating a sales and marketing force starting from nothing requires investors who "get it" and who are willing to invest to create this. In many wholly university-based entrepreneurial programs, the market force may be the collaborating companies and institutions who take the technology to the actual market, whatever this is. For academic/commercial developments of disruptive technologies, it is important to understand that the total answer to the sales and marketing needs cannot usually be distributors. Distributors will take orders, but they seldom, if ever, actively sell products for anyone, and, if they do, it will be for a large account where they can sell numerous units wherever they stop.

Finally, it must be understood that revenues of any new product will not be significant at the time when the rights to intellectual property or any other technological resource are acquired. Both of the valleys of death have to be crossed before revenues become significant (Fig. 2.1). This will be true for any technology, regardless of whether is adaptive or disruptive, purely academic, or an academic/commercial combination. It simply takes time for testing and evaluation to develop markets to reach mainstream users. Potential customers, especially larger mainstream ones, will be skeptical. Frequently, market success follows adoption by a few larger bell weather customers.

Funding across these valleys of death, so the technology or company does not die, is a formidable challenge. For purely academic research, the funding needs to come from granting agencies and/or partner organizations, while for academic/commercial projects, those two sources plus investment from whatever source are essential.

References

Allen, T., & O'Shea, R. (2007). *Paper series on academic entrepreneurship.* http://esd.mitedu/Headline/academicentrepreneurship.htm

Anonymous. (2009). *AUTM U.S. Licensing Activity Survey FY 2009.* AUTM (www.autm.net), Deerfield, IL.

Cohen, W. (2000). Taking care of business. *ASEE Prism Online*, 1–5.

Drucker, P. F. (1993). *Innovation and entrepreneurship.* Harper Business.

Goldman, M. I. (1984). Building a mecca for high technology. *Technology Review, 86*, 6–8.

Harman, G. E. (2019). 50 years of development of beneficial microbes for sustainable agriculture and society: Progress and challenges still to be met – Part of the solution to global warming and "hothouse earth". In P. Sing, V. J. Gupta, & P. Ratna (Eds.), *Interventions in agriculture and the environment.* Springer.

Kawahahi G. 2004. The Art of the Start: Penguin.

Livingston, J. (2007). *Founders at work. Stories of startups early days.* Apress.

McCouch, S. (2013). Feeding the future. *Nature, 499*, 23–24.

Moore, G. A. (1991). *Crossing the chasm.* HarperCollins.

Shane, S. (2004). *Academic entrepreneurship. University Spinoff and wealth creation.* Edward Elgar.

Shane, S. (2010). *The illusions of entrepreneurship: The costly myths that entrepreneurs, investors, and policy makers live by.*

Chapter 3
Innovations in Biology

Biolistics, Inc. and Sanford Scientific

In many cases, entrepreneurial scientists in academe and elsewhere deliberately set out to follow an uncommon path. In the 1980s, John Sanford, an associate professor at Cornell, focused on such a venture. His position assignment was to produce better varieties of small fruits such as raspberries and strawberries. At that time, plant genetic engineering (in this case limited to the insertion of novel genes that enhance some functions or properties) was just beginning.

Sanford could envision many ways to improve the plants were his area of responsibility through genetic engineering, but small fruits, like most other crops, were resistant to the then-available methods of genetic engineering (hereafter designated as plant genetic transformation). The issue was in the method of delivering foreign genetic materials into the plants. Existing methods relied on infection of plants by a bacterium that was then able to transfer a bit of genetic material that it was designed to carry into the plant. However, the bacterium only infected certain plants. Sanford thought about what else could be done to achieve this objective that did not involve bacterial infection and that could be a universal transformation for plants and for other organisms up to and including humans. He began to consider physical methods of introducing DNA into plant cells.

He examined several different concepts. One of the first was to use a microbeam to incise the wall of individual pollen grains and then to add the DNA to change the genome to the cut surface. This method was tedious and proved to be too injurious to the pollen grains to be useful.

Of course, teamwork is essential for progress. The most effective teams bring together a number of persons with similar goals but disparate skill sets and ways of thinking. John began to collaborate with Ed Wolf, who was a Cornell professor at the time specializing in energy beams and the like. They originally considered using a high-energy particle beam, but this too was very damaging to plant cells.

G. E. Harman, *Academic Entrepreneurship*,
https://doi.org/10.1007/978-3-031-06821-8_3

They next considered how fast a particle would need to travel in order to penetrate a plant cell. Their conclusion was that it was not very fast at all, perhaps as fast as a BB gun. They also need very small, but dense, particles to provide a few microholes in the plant cell wall. The research team coated 1–2 μm tungsten or gold beads with DNA that contained the reporter genes which can be used to identify transformation of plant cells. The DNA was mixed with protective substances, and thousands of these coated particles were added to the face of a cylindrical microprojectile. The microprojectile was fired and stopped by a plate with a hole smaller than the projectile. This forcible stopping ejected the tungsten or gold projectiles and impacted plant tissues a short distance away.

The original system was replaced by a system which used pressurized helium as the propellant, and the resulting particle gun was in a vacuum. The helium system provided an inert gas and allowed for programmable forces, so the acceleration was controllable. In this way, fine-tuning of the force of the discharge was possible to optimize for specific tissues and to provide maximum levels of transformation while minimizing cell damage. The vacuum avoided air friction that made projectile speeds variable.

The resulting device was the first biolistic transformation device, more commonly called the gene gun. This development was protected by series of patents owned by the Cornell Research Foundation, and Sanford and Wolf formed a company, Biolistics, Inc., that licensed the technology from Cornell. They contracted for the production of a few gene guns from a local manufacturer and sold these to some of the largest AgBiotech companies.

As the business began to grow, Sanford and Wolf decided their best option, rather than becoming gene gun manufacturers themselves was to sell the company, together with the IP backing it. The innovation was sold to Pioneer Hy-Bred International (now a part of DuPont) for several million dollars. Both Sanford and Wolf soon thereafter retired, and Sanford began another new business.

This company, Sanford Scientific, was set up a short distance from his former place of employment (Cornell University's campus at Geneva, NY). When Sanford sold Biolistics, he retained the license to the transformation system for use in ornamentals and turf. He began development of a variety of ornamentals and turf that would be resistant to diseases or offer other advantages. After a period of time, he entered into a partnership with the Scotts lawn and garden products company and began to develop transgenic grasses. This turf section was sold to Scotts and again John had a large payday.

After about 9 years, it became apparent to university researchers that the opinions of much of the public were so antagonistic to genetic engineering that it was fruitless to proceed. The author of this book had a similar experience and, like Sanford, abandoned the genetic engineering approach.

This anti-GMO opinion believed that genetic engineering was always dangerous and deleterious. Regardless of the product eventually produced, the method was so tarnished that this technology became limited mostly to academic research purposes and smaller entrepreneurs. Consequently, this area became the sole province of the very large companies, such as Monsanto. The public antagonism did not halt the use of GM projects. About 80% of the maize grown in the USA contains either transgenes, such as those conferring herbicide or insect resistance, or other genetic traits owned by Monsanto. In many cases, this also permits the company to sell chemical herbicides or other products specifically effective with the genetically transformed cultivar.

This is highly unfortunate as there are many situations where genes that are available for use with transgenic plants would make the plants much less susceptible to diseases and pests, increase plant productivity, or provide other benefits that would contribute to the evergreen revolution that is necessary now. However, the upshot was that academic and other smaller players were unable to enter and compete in these markets or participate in any way because the regulatory and opinion climate became so negative that only organizations with extremely deep pockets could afford to participate. Thus, the anti-GMO opinion resulted in the opposite of what was intended—it did not stop GMOs, and it gave the large companies a high degree of control over the seeds of important and critical food crops since they were the only ones with the financial strength to operate in the anti-GMO climate.

In the USA, the percentage of soybeans, cotton, and maize that are transgenic ranges now from about 70 to 95% of the total acreage planted to these crops (http://www.ers.usda.gov/Data/BiotechCrops/). Moreover, it gave the large companies almost monopolistic powers over crop production. One small example is that university researchers cannot conduct comparative tests on varieties any longer because of the large companies' restrictive regulations on use of their seeds. In many cases, academic researchers or other scientists can provide disinterested comparisons and chart alternatives to large company domination, but that is lacking in this situation with respect to GMO crops.

The Big Ideas Behind Biolistics and Sanford Scientific

1. Plants and other organisms can be improved using the genes that they already contain, and there are tremendous gains to be made in terms of resistance to disease (reduced pesticide use) or plant growth and development (greater yields with lower inputs) using genes that can be identified from other organisms.
2. In the mid-1980s, the methods of introducing genes into plants or other organisms were limited. If a universal method of gene introduction could be developed, it would be valuable and permit what was then perceived to be a tremendous opportunity to achieve the improvements that would be required for the evergreen revolution to proceed.
3. There were large opportunities also to use genetic engineering to improve plants that are used in amenity plantings—e.g., turf and ornamentals. This, at the time that Sanford Scientific was established, was considered an advantage for the smaller academic entrepreneur, since it was perceived that there would be fewer objections to GMOs in plants that are not used for food.

Lessons from Biolistics Inc. and Sanford Scientific

1. There is space for academic entrepreneurs with good ideas, even when large corporations are spending large sums of money in this area of innovation. The Biolistics example is excellent—the big corporations were indeed spending very large sums of money to produce universal transformation systems. However, Sanford and his colleagues approached the problem from a very nontraditional direction and succeeded where millions of large-corporation dollars did not.
2. If the idea is good enough and meets a strongly felt need, then it is possible to convert a good idea, through a small company, into considerable amounts of cash, very quickly. Biolistics existed for less than a year before it was purchased by Pioneer/DuPont. Both the university and the company owners profited almost immediately and handsomely.

   Biolistics, in particular, is an exception to the rule that it usually takes several years for academic research to be developed to the point where it becomes highly profitable and can generate high levels of revenues. The other start-ups in this section all took many years to be successful.
4. Even good ideas that can be profitable can be derailed by societal or financial events that are beyond the academic entrepreneur's control or that of his partners. The case of the unanticipated backlash against genetic engineering is a case in point.
5. Timing is everything. The Biolistics technologies came along at just the time when large companies needed a good method to transform plants, and Sanford and his colleagues sold their company at just the right time. The same was true with Sanford Scientific—if the sale to Scotts had been delayed even by a year or two, the transgenic climate would have been so negative that the deal would have been much less lucrative.
6. A strong IP position was imperative. While not emphasized in the narrative thus far, the fact that Sanford and his colleagues had a strong set of patents (at least six patents) owned by Cornell and available for license made the sales of Biolistics Inc. and Sanford Scientific possible. The biggest asset of the companies was their intellectual property.

CRISPR

CRISPR (clustered regularly interspaced palindromic repeats) is a tool for genetic alteration of any living organism. It is an example of the value of basic academic research in cooperation with company scientists.

One of the most important tools used in CRISPR technology is the gene *cas9*. This gene was isolated from *Streptococcus thermophilus*, which is used to produce dairy products (3). Bacteria are frequently infected with phages (bacterial viruses). The genome of these bacteria contains CRISPR sequences. Bacteria have evolved numerous natural defense mechanisms that target diverse steps of the phage life cycle, notably blocking absorption, preventing DNA injection, and restricting the incoming DNA. After virus challenge, the bacteria integrate genetic sequences derived from the phage genome. Plasmid DNA Cas9 performs this interrogation by unwinding foreign DNA and checking for sites complementary to the 20 base pair spacer region of the guide RNA. If the DNA substrate is complementary to the guide RNA, Cas9 cleaves the invading RNA and CRISPR, together with associated *cas* genes, and resistance specificity is determined by similarity to specific phage genes (3). DNA cleavage interferes with viral replication and thereby provides immunity to the host.

CRISPR is a "genetic scissors" to modify genes. It has two major ingredients: a *cas* gene and a small RNA molecule to a specific sequence of DNA that cuts the gene at the desired location. This allows genes to be modified, inactivated, or altered to obtain an altered phenotype (4). It is not the purpose of this chapter to describe all of the mechanisms involved, but this can be found in many references, including https://en.wikipedia.org/wiki/CRISPR_gene_editing (5).

There are many uses of the technology. For example, corn was modified by altering the *waxy* allele in inbred lines. Hybrids were agronomically superior and produced greater yields (6). In human medicine, CRISPR technologies have potential to treat a wide range of diseases, including cancer, progeria, heart disease, hemophilia, cystic fibrosis, and Huntington's disease and avoiding transplant rejection. It has been used to cure malaria in mosquitos, which could eliminate the vector and the human disease (5). CRISPR was used to edit the genome of the edible mushroom *Pleurotus eryngii* (7). Earlier, it was used to increase of resistance of bacteria to phages (8).

CRISPR is a powerful technology, and there is significant intellectual property. A search of patents reveals that there are about 700 patents filed or issued. One important patent was filed by the Broad Institute and Massachusetts Institute of Technology (9) and another is (10) filed by the University of California and the University of Vienna. In 2017, the patent office ruled that the Broad Institute, in affiliation with M.I.T. and Harvard, has rights to use CRISPR to modify DNA in the cells of humans, animals, and plants (11). However, anyone wanting to use the technology will probably need licenses from both the Broad Institute and the University of California (11).

This transformative technology has received numerous prizes and awards, including the Nobel Prize in chemistry which was awarded to Emmanuelle Charpentier and Jennifer A. Doudna for the development of a method for gene editing. In 2016, they also won the Tang Prize in Biomedical Science.

There are ethical concerns regarding such a powerful technology. While it can provide such advantages as curing genetic diseases, it also can be used to enhance qualities such as beauty, eye color, or intelligence and these are paths that ethicists should never take. Another avenue that should not be pursued is the editing of germ lines since this would result in a permanent change in the genome for future generations (12). As an example, in China, He Jianku altered DNA in human embryos and is now serving a prison term (13). However, it is relatively easy to do so anyone can edit genes for any trait they wish (13).

CRISPR is a good example of academic entrepreneurship. Doudna and her colleagues not only performed academic research, but they were also interested in commercial activities and in 2011 established Caribou, which was a company to bring their technologies to the market (13).

The Big Ideas from CRISPR

- Basic research toward applied goals can be used to develop breakthroughs in knowledge and applications that benefits society.
- These can be rewarding both socially and financially.

Lessons from CRISPR

- Such technologies can be universally applied to any living organism.
- Powerful technologies like CRISPR can be used for many purposes, and a strong ethical framework needs to be in place.
- Academic scientists can form companies—and in this case, this was encouraged by the university administration. This is a good example of the essentiality of a receptive culture within the university (Chap. 3).

Susan McCouch Rice Genetics Laboratory, Cornell University

Rice is a fundamental food crop for much of the world's population, along with maize, soybeans, wheat, and other staples. Most varieties are inbred, rather than hybrid, and none, or almost none, of the rice crop is genetically engineered. There are various reasons for this, including the fact that only about 1.5% of the global rice crop is produced in the USA (unlike maize and soybeans). It is also used primarily for human rather than animal feed or bioenergy; it is consumed largely as a whole grain, rather than being made into flour or other manufactured products; and 90% of rice is consumed within a few miles of where it was produced (14), so large and extensive distribution systems are not as important for rice as for the other crops just mentioned. All of this contributes to a local production model, and this has major implications for research programs that deal with this crop. The localized nature of rice production and consumption patterns makes it attractive to large AgBio companies.

Given that rice consumption is concentrated in regions of the world where population is growing the fastest, there is an urgent need to increase production to meet the local demand. However, given the localized distribution system and the fact that local varieties are generally preferred by consumers, there is little incentive for large AgBio to be involved. This is a place where academics and the universities need to take the lead. However, this must be done intelligently and with full utilization of the increasingly large data and tools available. Clearly plant breeding is a desirable avenue for this necessary improvement, but classical breeding systems primarily use domesticated varieties and locally adapted types as the starting point and only a very small fraction of the genetic diversity is accessible. The reasons for this are simple: as McCouch states:

> Plant breeders often worry that using wild species or unadapted landraces is too risky, scientifically and economically. It took 20 years and 34,000 attempts to cross a domesticated rice variety with a distantly related, highly salt-tolerant wild relative from India before fertile off-spring were obtained. It will take at least 4-5 years of breeding to eliminate unwanted wild characters (McCouch, 2013).

McCouch has a program to meet these needs. It is aggressively noncommercial. She states that "We work hard to partner with those who have a responsibility to feed hungry people. The role of rice research at a university like this is not about generating revenue—it's about the public mission of Cornell." However, the program is highly entrepreneurial according to the definition in the Preface, which means the transformation of an idea to an enterprise creates value—economic, social, cultural, or intellectual.

The classical approach that took 20 years and 34,000 attempts to produce improved varieties with enhanced characteristics from wild parents is too slow to meet the needs posed by increasing world populations, environmental pollution, and climate change.

One potent tool that can help improve the efficiency of breeding is that described in single nucleotide polymorphisms (SNPs). SNPs provide a diagnostic tool for identifying and monitoring regions of the plant's chromosomes that harbor genetic variation. A SNP marks a single nucleotide in the plant's DNA where a change (mutation) has occurred, one nucleotide out of the total of 400 million or so nucleotides in the rice genome. In cooperation with Affymetrix Corp., a SNP chip capable of monitoring one million SNP variations simultaneously across the 12 chromosomes of rice has been developed by the McCouch lab. The SNPs interrogated by the chip occur at specific known locations among the 400 million base pairs of the rice genome, and because their locations are known, they can be used to track the inheritance of specific genes or chromosomal regions that confer traits of interest in offspring that are derived from crosses between strains of rice that contain the SNPs. Thus, if nucleotide changes, detected as SNPs, occurring in specific genes of a wild relative are known to confer desired properties, the SNP signatures can be used to determine which of the offspring from an interspecific cross carry the desired gene or group of genes from the wild relative that confer the desirable trait. Young seedlings can then be rapidly screened by extracting a small amount of DNA from a portion of a young leaf and using the DNA sample to detect the individual's SNP signature. The SNP signature can be used to determine whether the individual carries the desirable trait(s) or the undesirable alternative(s), rather than having to laboriously wait to measure the trait(s) in mature plants in the field. Thus, it was possible for McCouch and her collaborators to identify genes that enhanced grain yield coming from wild relatives and to incorporate them into cultivated varieties. The robust new, high-yielding lines had sturdy stems that withstood hurricane winds plus provided a 20% yield increase. Similar results were obtained with collaborators in countries such as China, Indonesia, Brazil, Korea, and Sierra Leon (Ramanujan, 2011).

The McCouch research program develops genetic tools, phenotyping software, information systems, and biological materials that are available to researchers around the world. In addition to the SNP chips and datasets, other materials are available. For example, in collaboration with other researchers, they have developed RiceCyc, an interactive web-based database in which different sequences, genes, metabolites, or enzymes can be entered and biochemical pathways and interactions can be identified. Of great importance today, new strategies for evaluating the phenotype of a plant under diverse sets of conditions (i.e., resistance to pests, nutrient uptake, rate of growth, grain yield under drought) need to be developed. To be efficient, high-throughput systems need to be developed to provide accurate ways of assessing phenotypic variation and relating it to whole-plant performance in the field. In collaboration with USDA colleagues, they also have developed digital imaging systems and computer programs capable of analyzing the morphology and 3D architecture of plant root systems, which is a critically important aspect of plant performance. However, this phenotypic character has traditionally been difficult to measure and evaluate, in part because of the difficulty of removing intact roots from soil systems. The McCouch, MacCurdy, and Kochian groups have collaboratively developed imaging and computer-based analytical tools for this purpose (Clark et al., 2011, 2013). These tools assist in the evaluation of root growth among diverse plant varieties. The RiceCyc database and the software for measuring 2D and 3D root systems are open source and free to anyone who wishes to use them and are readily available on the web.

Thus, the McCouch program is providing a suite of useful tools for others to use for the genetic evaluation and improvement of rice, and she has also developed improved varieties of rice directly with cooperators. However, her role is not to directly produce finished varieties of rice but to give collaborators and breeders around the world "the genomic tools, strategies and enhanced varieties they will need to ensure an abundance of this essential crop" (Ramanujan, 2011). This program is dependent upon a high level of funding from public sources and private, nonprofit foundations. The program also depends upon cooperators around the world and at Cornell. The teams assembled must include plant breeders, field biologists, genomicists, computational biologists, and others assembled into interdisciplinary networks. These networks must be able to interact in beneficial ways to enhance rice productivity and sustainability to feed the world and to meet the environmental challenges of a growing and changing world.

This program, therefore, provides an entrepreneurial answer to a major world need—providing adequate food to a growing population. It is, or almost totally, funded through large grants from national and private, nonprofit organizations.

The Big Ideas Behind the Rice Project

1. Rice is among the most important crops in the world and is produced and consumed locally, with a large percentage of the world's population dependent upon this crop.
2. The production of rice needs to increase dramatically to meet world needs. The crop also needs to be able to be productive in the face of increasing environmental issues, including global warming, increased salinization, and other threats.
3. The total genetic resources of rice (and most other crops) are not well exploited. Much of the diversity and useful genes are in wild species that are not immediately useful in rice production.
4. Useful genes and traits can be identified in wild species, but standard plant breeding strategies require scores of years to incorporate them into useful, high-yielding varieties.
5. The advances in genotyping and phenotyping (analysis of the genes and the expression of the genes in plants) allow rapid introgression and selection of new varieties with commercial characteristics but with highly valuable yield and environmental resistance traits.
6. Advances in genomics and information technology make it possible to develop tools and share information with field scientists and breeders around the world at no or nominal costs.

Lessons from the Rice Project

1. The rice project demonstrates that it is possible to have a highly entrepreneurial program supported entirely on public, competitive grants, and private, not-for-profit foundation funding, with no profit motive anywhere in the university program.
2. This project was successful because:

    1. It is headed by an exceptional scientist who has vision and can assemble scientific and financial resources to accomplish this very large task.
    2. It deals with a crop that is critical to alleviation of world hunger, such a comprehensive system could not have been developed around a smaller, less important crop.
    3. It is an adaptive technology in that no basic new product had to be developed—improved rice will flow through existing channels and so totally new markets do not have to be developed. Thus, this is an adaptive and not a disruptive technology.
    4. The program was driven by a highly motivated and very capable leader.
    5. High levels of funding were obtained from a variety of sources, based on the demonstrated productivity of the program itself.

6. Because of the funding levels, the high regard internationally for the program, and the importance of the crop, the program attracted (and in many cases supported) a wide range of capabilities, from basic science to field application. The team that has been assembled is essential for the success of the program.
7. The program generates great societal value in terms of feeding a hungry world.
8. It also generates a very substantial amount of overhead that flows into the university, but it does not generate royalty income.
9. It is unlikely that Cornell has any solid numbers on the value of the project/technology systems, although, for example, the value of the increased value of rice produced could be estimated.

The System of Rice Intensification

The system of rice intensification (SRI) is a plant management system that avoids some of pitfalls of flooded rice culture. SRI involves alternate wetting and drying of rice paddies rather than continual flooding. Flooding is used to control weeds, but in SRI a rotary weeder serves this function. It enhances soil nutrients through application of organic matter rather than commercial fertilizers (15). It is used by about 20 million farmers (16).

One undesirable consequence of flooding is the production of methane by methanogenic archaea under the anaerobic conditions of this cultural method. Methane is an important greenhouse gas, so this is an environmental benefit afforded by SRI (17, 18).

SRI has been combined with a strain of *Trichoderma asperellum.* Combining the fungus with SRI has increased yields, greater water use efficiency, and greater photosynthetic rates (19, 20).

The Big Ideas Underlying SRI

1. The world needs enhanced food production, and systemic approaches such as SRI can provide this.

Lessons from SRI

1. Academic research can provide important advantages for farmers around the globe.
2. These programs can also provide significant environmental benefits.

3. Very importantly, it emphasizes the importance of system approaches. No technology, however useful, can confer advantages by itself.

Broccoli Project (No New Company): An Academic-Private Partnership

Not all university entrepreneurial activities require the formation of a new company. In fact, in agriculturally based technologies, most do not. A very important reason why a new company may not be a requirement is because the entrepreneurial project deals with an existing salable product. An important advantage of such projects is that market channels already exist and there already exist. A new and improved existing product does not have the long period of product development and product that make a new company developing a new and untried technology or product so difficult. The time to market is therefore much shorter.

An example is the broccoli project. This project had its genesis in the concept that almost all broccolis sold in the USA is grown in California or outside the country and thus has to be shipped a considerable distance.

This limits profit potential for eastern growers and also increases costs to customers because of the long shipping distances required. The reason why this situation occurs is biological—if broccoli is grown under hot conditions (maximum low night temperatures of about 22–24 °C), then the flower buds do not develop and the heads are very uneven and distorted, making them unsightly and unsalable. The conditions in California in the winter are appropriate for producing attractive produce, but in the east in the summer, temperatures are too high and, in the winter, clearly, temperatures are too cold. Consequently, there are almost no commercial east coast broccoli production.

However, biological solutions to the problem exist. The university and private scientists who are engaged in plant breeding have 20 or more years of experience in producing east coast-adapted broccoli, but there has been no infrastructure to create a market. Specifically, no seed companies wanted to license the east coast-adapted varieties that were developed because there was no market for the seeds.

Thus, there was a chicken and egg situation. There was no market for licensing varieties because there was no commercial east coast broccoli industry. The need, therefore, was to create the necessary infrastructure, including appropriate broccoli varieties, growers who would produce them, and industries or companies that would purchase the product, assuming that the quality was appropriate.

Associate Professor Thomas Bjorkman decided to attempt to create a system that combined all the necessary components of a new industry. He discussed the issues with growers, plant breeders, and commercial markets, such as supermarkets, to try to find a solution. The greatest problem was that, to put an integrated system in place, it required a considerable amount of funding. The potential participants all liked the concept, but none wished to risk time, money, and effort on an incomplete system.

Recently, the USDA has developed a Specialty Crops Research Initiative funding program. Dr. Bjorkman applied for and received a planning grant for the project. He convened a 2-day meeting of growers, university extension personnel, breeders, distributors of produce, and final marketers such as supermarket chains. The participants came from all over the eastern USA, and public employees included USDA and university personnel.

Since this was a publicly funded project, no potential participants were excluded. However, any intellectual property belonging to universities or private companies, especially seed companies, had to be protected. This includes primarily the breeding lines and other plant materials. In most cases since markets and varietal choices are unknown, these are not covered by formal intellectual property filings but instead are proprietary to the university, seed companies, or other organizations, including varieties submitted by federal scientists in the USDA. If there was no protection, then no one who has proprietary lines would allow them to be included in the project. However, there also was a requirement that trials be conducted so that varieties could be compared side-by-side in different growing and environmental circumstances. The answer to this dilemma is that any lines or varieties included are provided to organizers who give each individual line a coded entry number. These entries are then placed in trials in several different locations across the eastern USA under the control of public scientists. Seeds and plants are protected against unauthorized acquisition or use of any biological materials. Anyone who is interested can view the trials and lines and varieties of special interest can be conveyed to the trial organizers. Beyond this, all the data on the trials is analyzed statistically and compiled and made publicly available. It is anticipated that any variety or line selected for commercialization would be grown by seed companies for sale, with appropriate royalty agreements to any public institution that develops the line.

Beyond this, advanced lines require development of specific cultural and harvest conditions to provide a quality product. This is responsibility primarily of public scientists, although private companies can do their own horticultural adaptation research for specific clients.

This all requires a substantial amount of financing. The USDA grant program requires matching funds from private companies to obtain governmental funds. For this purpose, $3.2 million was secured from private companies to pair with $3.2 million in USDA funds. The private funding was from a consortium of seed companies, producers, farmers and growers, and retailers such as supermarkets.

The project is fully up and running now. Final results are expected within a year. While data is available publicly, the participants have the inside track on selection and processes since they are fully engaged with the project.

This project is expected to provide income to the public institutions through royalties. Growers have new and lucrative markets for a high value product, to seed companies through new varieties to sell and to distributors and retailers through provision of an attractive local product that ought to be less expensive and of higher quality through a trucked-in product. It will create jobs, albeit in many cases at a low wage. For local governments, it provides new economic opportunities and a more prosperous local economy.

The Big Ideas Behind the Broccoli Project

- There is great potential in introducing new or unusual organisms into specific agricultural systems. In this case, a common vegetable, broccoli, could be placed into the eastern US agri-ecosystem if appropriate varieties and infrastructure were identified or created. This would provide fresher produce and reduce shipping, in addition to appealing to the locavores movement.
- There was a significant amount of intellectual capital from publicly funded work that was underutilized, especially in the 20+ year effort that has gone into breeding and culture of broccoli and other similar crops.
- To utilize this intellectual resource, seed companies, distributors, retailers, and marketers had to be brought together into a mutually beneficial relationship that could create financial value to all parties concerned.
- Intellectual and proprietary property, especially as related to the genetics and identity of candidate varieties, had to be protected.
- The USDA Specialty Crops Research Initiative promoted the required funding to pull all of the disparate groups together into a synergistic whole, and there had to be matching funds from potential users.

Lessons from the Broccoli Project

1. There are opportunities for academic entrepreneurs to make a big difference even without starting or involving a specific company. The opportunities are primarily in areas where there are existing market channels for the products being produced. If the technology being developed involves creation of wholly new products and markets, then a specific company to develop or license it is probably essential.
2. Academic entrepreneurs may be able to create a structure or an infrastructure that allows unused or underutilized public intellectual capital to reach the marketplace.
3. Such academic entrepreneurs ought to be encouraged to "think outside the box" and to determine for themselves how they can make such a difference. They ought to be rewarded for doing so even if the outcome is not a typical academic value—e.g., publications and the like. Such grassroot efforts ought to be highly prized.
4. The university IP position and capital were already available. The innovation in this case was devising a method and a structure whereby the technology and capital could be utilized.

The Trichoderma *Effort: Several Companies Have Been Involved*

Trichoderma spp. are fungi that are among the most common free-living organisms in soil. They are typically not pathogens and do not cause harm to living organism. They were discovered to have ability to parasitize or produce antibiotics against other harmful fungi even as long ago as the 1930s (Weindling, 1932, 1934). Ever since then, there has been a dream of creating nonpesticidal methods to control plant diseases. In the years between the 1930s and the mid-1990s, there were a great many publications on the topic, a great deal of information was gathered, and attempts to commercialize the fungi were initiated. However, they never became a significant tool in plant agriculture. One reason for this was timing during that period (before Rachel Carson's *Silent Spring),* agriculture was dominated by synthetic pesticides and fertilizers and there was little perception that a gentler, more environmentally friendly approach was desirable or potentially profitable to the large AgBio companies of the day.

The author began working with cooperators, especially Ilan Chet of the Hebrew University, in 1980. As more and more knowledge was gained, it became apparent that good results could only be achieved with improved and highly selected strains. This is clearly apparent since the fungi are very abundant in soil and so for them to have a measurable impact, the strains to be used in plant agriculture must be improved over the common ones.

In the mid-1980s, the author concluded that just doing research on these potentially useful fungi and publishing the results in the scientific literature was going to do little unless there was a concerted effort to develop products and specific uses for the fungi for commercial practice. Otherwise, the work he did and that of many others was "just academic" and could not benefit anyone, either as an aid to improve plant agriculture or to provide jobs or create wealth and jobs. For this effort to succeed, he and his colleagues had to find funding sources that were sympathetic to and supportive of research that attempted to cross the "valleys of death" to create useful products. The expected funding sources such as federal government grants at that time were not geared to such efforts, but one was—this was the US-Israel Binational Agricultural Development fund, and he and Dr. Chet received numerous awards over the years, without which there would be no *Trichoderma* story. Another funding source that was critical was NYS Center for Advanced Studies program—this foresighted program provided grants specifically to develop technologies and products that would benefit the NYS economy. Such programs are essential in any university that attempts to create jobs and translate technology into useful products.

In the mid-1980s, he and his colleagues produced asexual hybrids of the fungi to create strains that were superior to others that were available. The fungi lack a sexual stage, so asexual hybridization was the only route to combine characteristics of different strains. This effort was successful. Thousands of strains were produced and screened, and a few stood out as being highly improved. A major attribute of the new strains was a strong ability to be "rhizosphere competent," i.e., to colonize roots from a seed treatment, to grow in the root, and to colonize it for the life of at least an annual crop. Other strains typically colonize roots, but they do not continue to grow and become quiescent and so do not provide long-term plant benefits. The best early strain was *T. harzianum* strain T22.

This strain and the IP behind it were licensed to the Eastman Kodak Company from the Cornell patent office in about 1987. It was to be the flagship product for the company in AgBiotech. Kodak spent several million dollars over about the next 3 years on the organism. Since it was to be used as biocontrol agent against plant pathogenic fungi, an EPA registration was required before it could be sold. One of the very useful things that Kodak did was to do the toxicology and to obtain the registration. Other results were less good. First, at the time that Kodak acquired the technology, the unit of production was a petri dish. Until a useful method of production is developed and a product is produced, no sales are possible. Kodak had a large fermentation facility in Rochester and part of the reason for licensing T22 was to have a use for the facility. Unfortunately, the methods of liquid fermentation they developed produced high levels of biomass, but the shelf life was abysmally short and noncommercial. Nonetheless, they proceeded with field trials as a seed treatment, and results were poor. It was never possible to ascertain whether the poor results were due to the potential product itself of was just because the organism did not survive long enough for the trials to be completed. After about 3 years of development, Kodak experienced the Polaroid lawsuit judgment against them and they abandoned all of their non-imaging technologies, including T22. After an attempt to find a buyer for the technology, they transferred the technology back to Cornell, including the EPA registration package. The author and two colleagues decided that they had a great opportunity to develop the technology themselves. They formed a company, TGT Inc. (now BioWorks Inc.), and obtained rights from the Cornell technology licensing office. The fledgling company did have a better handle on how to produce *Trichoderma* more economically and that had adequate shelf life, but the founders were totally naive in business and still lacked many sorts of information as to how to use the fungal products most appropriately to best advantage. One large opportunity and a huge disadvantage was that the fungi conferred many advantages to plants to which they were applied; they increased plant growth and controlled diseases. Which attribute ought the new company pursue? In addition, there were many potential markets—seed treatments, greenhouse soil amendments, and application as granules—and it was not clear at all where the markets were or where the products might be most successful. Typically, start-ups try to cover too much, but in this case, where it was unclear where the best fit of the products and fungi were, it was important to cast a wide net.

An early priority had to be fundraising. In most cases, new start-ups can sell stock only to qualified investors, which are defined as persons with $one million in net worth or who earn over $250 K per year. However, there is a one-time opportunity to sell shares to anyone on a section 504 exemption under article D of the federal securities law. Given this, and having little other option, we set out to sell to friends and neighbors. This is a typical situation—at the start of a new enterprise, usually only the 3Fs (fools, friends, and family) will buy into such an untried enterprise (Kawasaki, 2004). So, in a feat that still seems remarkable to the author, the three cofounders set out to, and did, raise $one million from these unqualified investors. The only thing other than unproven IP that the company has to sell is the credibility of themselves. So, in such a venture, the company founders were asking small investors to provide money based primarily on their view of the founders. It is, and was, an uncomfortable situation and not one the author would repeat.

An early event in the company was an almost immediate order for a sizable amount of product by a large agricultural distribution company. There were not established production facilities, but the founders of the company themselves, in crude and difficult circumstances, managed to fulfill the order with product of reasonable quality. We congratulated ourselves on this accomplishment and sat back to wait for further orders, and waited and waited, while in the meantime the small amount of capital of TGT dwindled away. The next orders never came. We eventually visited distributor warehouses and found all of the product unopened and unused. In conversations with the warehouse managers, we were told that the product was delivered to them by the company, but they had no idea how to use it, so they did not. In retrospect, this was a huge error on our part. As soon as the product was delivered, we needed to be working directly with the distributor warehouses on what the product was and to set up an aggressive program of product placement and evaluation. This is essential with a new product. The situation became worse over time—after we discovered the problem, we wanted to do the placement and evaluation, but by that time, the product shelf life had expired. TGT could not afford to take back and replace the product, so the company sold it anyway. We pointed out that this was contrary to our contract, and the company said, "so sue us." Of course, we could not do that either because we had insufficient funds to do so. This points up a serious problem with contracts between large and small companies—early-stage companies are typically underfunded the financial capability to protect their contracts or their intellectual property.

However, over time, sales did begin, primarily in the seed treatment and greenhouse markets. There was a need, as there almost always is, to improve the production facility to make it more efficient and capable of producing higher quality products. Therefore, TGT (which had then changed its name to BioWorks) embarked on another fund raise of $3.5 million. This also was successful. It should be noted that not only did we have the organisms themselves, but we also had cloned several genes and inserted them into plants where they induced a high amount of resistance to plant diseases. Thus, there were two strong components of the company.

The new production facility was developed, and, to the dismay of the founders and investors, it proved to have unforeseen problems. This, which occurred just after the facility was opened and the fundraising round completed, was of great concern. The biggest problem turned out to be totally unexpected—we had UV lights for sterilization of air over the top of a plenum, so the light never shown on the growing organism, but we discovered, finally, that reflection of the UV rays was damaging growth of the organism. This was totally unobvious but was corrected, and this same basic design is being widely used now. It should also be noted that the initial production levels were about 5×10^8 colony-forming units of final product, and the levels now are almost 100x greater. This is a huge improvement that makes the cost of production much lower than in these early days. This improvement was not any single large factor but just the accumulation of many small ones. Continual improvement in manufacturing is critically important.

By this time, the total number of employees was about ten, and of course, this increases the burn rate. A series of factors then occurred that resulted in great dissension between the Board, the founders, and the founders themselves. There is a great tendency if things start being difficult for everyone to try to do everyone's job. There is a fine line between solving problems and getting in everyone's way, and it is a very good way to lose friends. At the end, the two cofounders were asked to leave the company, and after a short period of time the author became the acting CEO. It was essential at that time for him to reduce his employment at Cornell and for about 2 years, he split his time between Cornell and BioWorks. This was a productive and a great learning experience for him, and he could have stayed as the CEO and left Cornell entirely. The temptation was great, but in the end, the lure of tenure won out.

A new CEO was hired, and the viewpoints of the author and the new CEO differed sharply. Sales at the time of hiring were increasing, but of course not as fast as anyone hoped. The new CEO changed entirely the sales operation of BioWorks and, unfortunately, this was a retrogressive step that resulted in sales declines. Further, the new CEO decided to focus on a single area of sales, the greenhouse soil disease control market. The marketing ship was eventually righted, and this, over several years, became quite successful. BioWorks now enjoys good markets in the USA and around the world. Now, anywhere the author goes to greenhouse operators and asks if they know about T22, RootShield and PlantShield, the answer is invariably yes, and all of them believe that the products are really good. When they understand that the author is the inventor, they are uniformly appreciative.

In addition, in the mid-1990s, the author was still employed part time at BioWorks and endeavoring to develop the transgene systems for licensing to large companies. The introduction to this chapter describes how the transgene biotech industry developed and the reasons why small firms could not compete. This was a significant blow to BioWorks, and one that, at the time the Company was founded, could not have been foreseen. Because of the slow growth of sales and because the small company transgene company markets became impossible, BioWorks had to raise money twice more, and the last one was disastrous to the founders' and early shareholder's share value. If survival money is required, it becomes very expensive if it is possible to raise funds at all. The last fund raise was about $0.5 million of a total of about $six million raised, and there was a 20:1 preference of the new series over the founders or the first series.

This was coupled with the fact that the focus on the greenhouse market was directed to a total market size of only about $50 million, so if BioWorks gained even a 20% market share, the total sales revenue was not large. Management initiated a "cram-down" based on the voting rights of the 20:1 preference. The author attempted to stem this tide by finding other investors but was unsuccessful and, in the end, management was successful. This meant that the founders and the early investors lost their time and money, and this was a great shame.

However, in many respects, BioWorks was and is a success story. It still exists, and while the author is no longer in touch with the company, it continues to exist in Victor, NY, and generate economic value for the region. Here are some of the successes and drawbacks:

1. T22 was the first reasonably success *Trichoderma* product that was marketed anywhere in the world. Other products came before and at about the same time as T22, but they had, at best, limited success. There are more than 100*Trichoderma*-based products in the world today (Woo et al., 1991) but as the author is told by many persons around the world, without the success of T22 leading the way, commercial success, and the credibility for the whole field that was generated, would have been very difficult.
2. This commercialization effort was critical to the author's academic success in his field. He probably is the world leader in this technology and is invited to give invited lectures and write invited papers in prestigious journals, and the commercialization success was a principal factor in this recognition and success. The commercial efforts and data were highly synergistic with the basic program he developed. He is recognized both for his pioneering efforts in the science of beneficial plant-microbe interactions and as the leader in commercialization for microbial plant inoculants.
3. T22 has been good for Cornell and has generated more than $two million in royalty income. This is among the highest for any technology in the College of Agriculture and Life Sciences at Cornell University.
4. However, it was only a modest success—the company did not grow and prosper as it could have.

Of course, the biggest difficulty was the fact many of the investors in BioWorks, due to the series of funding rounds that were required, and the management cram-down, lost their investment. But that was not the end of the story by any means. In the year 2000, a company was established in Van Wert, Ohio, by a group of persons from the ag industry. This company was founded primarily to commercialize technology from the author's research and that of another scientist, David Kuykendall. The primary focus was on field crop application of *Trichoderma* strains and *Bradyrhizobium* (soybean nitrogen-fixing bacteria). As mentioned earlier, BioWorks management decided not to proceed with field crop applications and so provided a license to ABM to commercialize T22 and other strains on maize, soybeans, and other field crops. Earlier, cooperative research between BioWorks and ABM had demonstrated that, in some cases, there were remarkable increases in maize yield and productivity (Harman, 2000). Not only did T22 increase yields and growth of maize, but it also was found to improve the crop's ability to use nitrogen use efficiently. ABM therefore set up an aggressive field trial program with T22 as a seed treatment on maize. Unfortunately, after several hundred trials were completed (which would be impossible from a solely university base), it was found that T22 gave positive yield responses most of the time, but that with some maize lines, it gave a negative response. This precluded its use as a biological seed treatment on maize although it worked quite well on wheat (21). Many researchers had discovered that *Trichoderma* strains sometimes increased plant growth, but the mechanisms were a mystery. However, at about that time, –omic technologies became available (e.g., genomics, proteomics, and metabolomics). Researchers around the world embraced these new tools and we began finally to understand how *Trichoderma* really worked. At about the year 2000, *Trichoderma* spp. were still considered to act primarily as antibiotic producers and mycoparasites, but some studies demonstrated that they induced systemic resistance to pathogens in plants (Bigirimana et al., 1997; Yedidia et al., 1999, 2000, 2003). Further, it was demonstrated the *T. asperellum* did not just colonize the exterior of roots, but it penetrated the cortical tissues, where it induced the plant to wall off hyphae of the strain in an asymptomatic infection (Yedidia et al., 1999). This finding was later extended to other fungi. This work was accompanied by a series of papers that demonstrated that the beneficial fungi dramatically altered the expression of proteins and genes in plants systemically (Djonovic et al., 2006; Marra et al., 2006; Alfano et al., 2007; Shoresh & Harman, 2008a). These findings were surprising because expression of a greater number of proteins or genes was differentially expressed in the shoots than in the roots, even though the strains were located totally in roots (Shoresh & Harman, 2008b). The ability of the fungi to change plant gene and protein expression are driven by specific elicitors (effectors) produced by different strains. From this conceptual framework, it was apparent the international *Trichoderma* community had, at best, an incomplete concept of the mechanisms by which their benefits occur and that, in fact, the fungi are best considered as endophytic plant symbionts (Harman et al., 2004). The elicitors range from hydrophobic proteins to small-molecular-weight materials, and it is anticipated that the nature of the plant-*Trichoderma* is largely controlled by the specific elicitors that are produced by the organism (Harman et al., 2004a, b; Shoresh et al., 2010). These interactions frequently give rise to larger plants (Fig. 3.1).

This dramatic revision of the plant-*Trichoderma* interaction has substantially changed our understanding of the capabilities of this system. It is much more versatile and much more useful than we believed. We now know that almost all the plant disease control of *Trichoderma* is provided by some form or other of induced systemic resistance, although the specific mechanism may differ from classical definitions of induced systemic resistance or systemic acquired resistance (Harman et al., 2004a, b; Marra et al., 2006; Bae et al., 2011). This clearly has been demonstrated even for "classical" systems where antibiosis or mycoparasitism was considered to the primary modes of action—genetic analysis clearly shows that even there, systemic resistance is the sole mode of action (Hanson & Howell, 2001; Shoresh et al., 2010). This discovery was of fundamental importance for two reasons: (1) the level of disease control by induced resistance is not absolute and is primarily effective only for its protective value and for protection against low levels of disease (Walters, 2010) and (2) if one tries to optimize biocontrol systems for the wrong mechanisms, failure is likely. Since products were optimized to improve mycoparasitism and antibiosis, the expected improvements in formation and delivery were not realized.

Fig. 3.1 Photo of corn plants grown from seed treated with *Trichoderma* (right) or nontreated (left). Plants are from a commercial field in Wisconsin

However, even the relatively low level of disease control is important especially since *Trichoderma* growing on roots can create long-term levels of resistance, and this is useful but is inadequate for some situations.

Undoubtedly more important is the other improvements that the systemic effect of the fungi can induce in plants. These include:

- Consistent and reliable improvements in crop yields of critically important food crops, including maize (Fig. 3.1), wheat, soybeans, and rice (Harman, 2011a, b)
- Improvements in the resistance of abiotic stresses such as temperature, drought and salinity (Mastouri et al., 2010; Shoresh et al., 2010), and polluted soils
- Improvements in the abilities of plants to take up and use nitrogen fertilizers more efficiently; plants typically take up about 33% of the nitrogen applied, and our data clearly demonstrates that we can reduce nitrogen fertilizer by about 50% with no reduction in crop yield. This effect is widespread—similar results were obtained in greenhouse chrysanthemums, wheat in upstate New York, and rice is Southeast Asia (Shoresh et al., 2010; Harman 2011a, b)
- Improved photosynthetic efficiency

These improvements in the knowledge base and the development of strains to take advantage of them create large new economic opportunities. Many crops other than maize can obtain similar advantages. Moreover, any delivery system that provides contact of the beneficial fungi with roots is effective. This makes possible novel new fertilizers that do not pollute waterways with nitrate that is probably the single most damaging agricultural input and many other products. The advantages to the developing world are very large, since *Trichoderma* can replace or enhance inputs necessary to high yields (Harman, 2011a, b). This makes these fungi very attractive internationally.

This project had another important transitional event as the author retired from Cornell and went to work full time as the Chief Scientific Officer at ABM. This transition, though largely driven by the lack of transparency in the COI process, has been very beneficial both to the author and to the company. It also will be beneficial to Cornell through royalty income.

Within the past year, the author and ABM have been in the process of establishing a state-of-the-art laboratory for genomic discovery and product development. It is staffed with qualified individuals including several PhDs that are company employees, and the lab is well equipped with the necessary tools. This includes state-of-the-art -omics equipment and more applied items such as greenhouses and fermentation/microbiological production tools. It is now obvious that it is possible to use genomic sequencing of specific microbes to make major advances. There is a significant amount of information regarding specific genes and gene cassettes that encode important metabolites that are involved in the symbiotic process. We have a wide variety of microbial diversity, and we will be able to link specific expressed microbial genes to particular capabilities to enhance plant performance. These will range from enhanced yields (synergistic mixtures of microbes will be important) to performance characteristics such as resistance to abiotic stresses, disease and pathogen control, increased nutrient use efficiency, improved photosynthetic capability, and others. This will allow us tailor organisms (through selection among naturally occurring strains) to enhance multiple crops and to provide specific advantages.

This revolution involved triggering expression of multiple genes and pathways in plants through interaction of microbial triggering metabolites with control and response elements of plants. Microbes have already evolved these triggering metabolites and we now need to understand and exploit these. This really is the linkage of microbial and plant genomics, expedited by the advances in rapid phenotypic analyses. This is already well underway and is resulting in substantial improvements in ABM's financial outlook. Earlier, we quoted from a press release from Novozymes and Monsanto that stated they wanted to create a "powerhouse with a unique opportunity and approach to unleash the transformational opportunity in naturally derived microbial solutions in agriculture." This is precisely the area in which the author's programs and companies are situated and have been in place for the past 20 years. The technology is now sufficiently mature so that it is possible to reach this goal. ABM is an entrepreneurial position that believes that in biological knowledge there is power for commercial success and to do good in the world. This, coupled with the new technologies in the various -omics fields, together with adequate funding, highly capable staff, and appropriate equipment (the core facilities at Cornell will be important) is permitting even our small company to compete with the larger companies. ABM is in discussions with a variety of companies that will permit rapid economic growth, each with a specific contribution to make. Because of confidentiality agreements, it is not possible to be more specific at this time. ABM and its R&D facility also have in place a network of highly qualified scientists worldwide who contribute. This is allowing ABM to occupy a significant space in this rapidly developing field.

On a personal note, all of this is very gratifying to the author. After decades of effort, the goals of a major role for selected and highly efficient microbes in commercial agriculture is becoming a reality. At the outset, in the BioWorks days, technological advances and equipment were not available to make the sorts of advances possible now with the -omics technologies. However, if BioWorks had not been established in 1993, and developed concepts and at least limited commercial success, ABM and the successes we are enjoying now could not have occurred.

Beyond this, clearly the successes of ABM are directly improving the author's financial status, but for most of the time between 1993 and about 2010, there was a negative effect of the companies on his financial well-being. There were royalties from Cornell, but the costs directly associated with the companies and their needs made the entire effort a net financial negative, sometimes with large amounts.

There is also the concept of establishing a legacy of a long career. If the author had worked only at Cornell, his academic successes would have been greatly lessened because the companies provided direct financial support to his program and because a number of initial important observations and discoveries were made in the field. It is also certain that regardless of success, the author's program would have ceased totally upon his retirement. His position at Cornell and his research program are not being continued. This is partially due to the poor financial climate of the institution, and because of his entrepreneurial role, he was outside the mainstream of the College of Agricultural and Life Sciences. However, his new role of full-time CSO of ABM means that he has hired an outstanding staff, including several PhDs, and one of these is being mentored to become the CSO when Harman finally retires. Thus, both the scientific program and the benefits to the world provided by his efforts will continue. This is a value beyond price to him.

The Big Ideas Behind the *Trichoderma* Project

1. Microbial agents such as fungi in the genus *Trichoderma* can function as alternatives to chemical pesticides.
2. The advent of genomics and related technologies create major new opportunities for highly advantageous products that have not heretofore been available.
3. The concept that selected strains of these fungi are multifunctional and provide multiple benefits to plants is a major step forward.

Lessons from the *Trichoderma* Project

1. Academic entrepreneurs, and the level of business people they can attract to a new business, may be naïve regarding the amount of money and time required to start an enterprise.
2. Technologies such as *Trichoderma* will be regarded with skepticism if it represents a paradigm shift in, especially, a conservative industry such as agriculture. Everyone is from Missouri, i.e., the products have to undergo an extensive period of development and testing before they will be accepted.
3. A new product may have many potential uses as was the case with this project. It takes time and money to identify and prove the products.
4. In short, the "valleys of death" were large and formidable in this case.
5. At the start, almost the only capital that a new company like TGT has is its IP and the reputation of its founders.
6. However, a university base for its technology founders is a huge advantage and one that can provide excellent return to the university in which it is located.
7. Grant programs to cross the "valleys of death" are absolutely required and essential. Few or no private investors will invest the level of funding required over several years.
8. Investment in early stages by companies is necessarily speculative and investors may lose their money although the potential for major gains is real. Both the founders and the investors must realize that this is the case.
9. However, once the products are established in the marketplace, they can provide paradigm shifts in the way that agriculture or other businesses operate.
10. Sudden breakthroughs like the ones based on genomics in this example can remarkably change the footprint and impact of the technologies. However, new start-ups, just like larger and established companies, have to be receptive to change and new developments and be willing to focus on and adopt them.
11. The opportunities open now can rapidly result in improved plant performance and are based upon an intimate understanding of the interaction of microbial genome components and their effects upon plant gene expression.
12. The maturation of the technologies can provide an excellent return on public funding.
13. As described in the Preface, the blend of good science and commercial success has immeasurably increased the reputation of the author worldwide.
14. It also contributes greatly to the societal impact of his program and is financially rewarding. The company and its direction also ensures that his legacy and benefits to society of his efforts will continue. Without the company, the program and its benefits would have ceased upon his retirement from Cornell.

Lessons from Chap. 3

1. There are many valid models for entrepreneurship in the academe. This chapter lists six very different ones.
2. The models that are most useful in any particular situation will depend on the type of technology that will be utilized, an investigation of the opportunities and possibilities afforded by the institution and the preferences and predilections of the technology inventor.
3. Three of the examples provided are adaptive and do not include a commercial entity, while the other three are disruptive and absolutely required establishing a commercial entity.
4. If a commercial entity is involved, the inventor probably needs to "have skin in the game," i.e., to invest personally in it. Otherwise, other funding sources will not take the opportunity seriously.

References

Alfano, G., Lewis Ivey, M. L., Cakir, C., Bos, J. I. B., Miller, S. A., Madden, L. V., Kamoun, S., & Hoitink, H. A. J. (2007). Systemic modulation of gene expression in tomato by *Trichoderma harzianum* 382. *Phytopathology, 97*, 429–437.

Bae, H., Roberts, D. P., Lim, H.-S., Strem, M., Park, S.-C., Ryu, C.-M., Melnick, R., & Bailey, B. A. (2011). Endophytic *Trichoderma* isolates from tropical environments delay disease and induce resistance against *Phytophthora capsici* in hot pepper using multiple mechanisms. *Molecular Plant Microbe Interactions, 24*, 336–351.

Bigirimana, J., De Meyer, G., Poppe, J., Elad, Y., & Hofte, M. (1997). Induction of systemic resistance on bean *(Phaseolus vulgaris)* by *Trichoderma harzianum. Med Fac Landbouww Univ Gent, 62*, 1001–1007.

Clark, R. T., MacCurdy, R. B., Jung, J. K., Schaff, J. E., McCouch, S. R., Aneshansley, D. J., & Kochian, L. V. (2011). Three-dimensional root phenotyping with novel imaging and software platform. *Plant Physiology, 156*, 455–465.

Clark, R. T., Famoso, A. N., Zhao, K., Schaff, J. E., Craft, E. J., Bustamante, C. D., McCouch, S. R., Aneshansley, D. J., & Kochian, L. V. (2013). High-throughput two dimensional root system phenotype platform facilitates genetic analysis of root growth and development. *Plant, Cell & Environment, 37*, 454–466.

Djonovic, S., Pozo, M. J., Dangott, L. J., Howell, C. R., & Kenerley, C. M. (2006). Sm1, a proteinaceous elicitor secreted by the biocontrol fungus *Trichoderma virens* induces plant defense responses and systemic resistance. *Molecular Plant Microbe Interactions, 8*, 838853.

Hanson, L. E., & Howell, C. R. (2001). Elicitor protein produced by *Trichoderma virens* that induces defense response in plants. US Patent 6,242,420, 1–14.

Harman, G. E. (2000). Myths and dogmas of biocontrol. Changes in perceptions derived from research on *Trichoderma harzianum* T-22. *Plant Disease, 84*, 377–393.

Harman, G. E. (2011a). Multifunctional fungal plant symbionts: New tools to enhance plant growth and productivity. *The New Phytologist, 189*, 647–649.

Harman, G. E. (2011b). *Trichoderma – Not* just for biocontrol anymore. *Phytoparasitica, 39*, 103108.

Harman, G. E., Howell, C. R., Viterbo, A., Chet, I., & Lorito, M. (2004a). *Trichoderma* species – Opportunistic, avirulent plant symbionts. *Nature Review and Microbiology, 2*, 43–56.
Harman, G. E., Petzoldt, R., Comis, A., & Chen, J. (2004b). Interactions between *Trichoderma harzianum* strain T22 and maize inbred line Mo17 and effects of these interactions on diseases caused by *Pythium ultimum* and *Colletotrichum graminicola. Phytopathology, 94*, 147–153.
Kawasaki, G. (2004). *The art of the start*. Penguin Books.
Marra, R., Ambrosino, P., Carbone, V., Vinale, F., Woo, S. L., Ruocco, M., Ciliento, R., Lanzuise, S., Ferraioli, S., Soriente, I., Turra, D., Fogliano, V., Scala, F., & Lorito, M. (2006). Study of the three-way interaction between *Trichoderma atroviride,* plant and fungal pathogens using a proteome approach. *Current Genetics, 50*, 307–321.
Mastouri, F., Bjorkman, T., & Harman, G. E. (2010). Seed treatment & with *Trichoderma harzianum* alleviate biotic, abiotic and physiological stresses in germinating seeds and seedlings. *Phytopathology, 100*, 1213–1221.
McCouch, S. R., et al. (2013). Feeding the future. *Nature, 499*, 23–24.
Ramanujan K. 2011. New tricks for a very old crop: Working across disciplines, rice researchers on campus are finding novel ways to head off global food shortages. Ezra Magazine, *Cornell University, III*, 10.
Shoresh M, Harman GE. 2008. Genome-wide identification, expression and chromosomal location of the chitinase genes in Zea mays. *Molec Gen Genom.280*, 173–185.
Shoresh M, Harman GE, Mastouri F. 2010. Induced Systemic Resistance and Plant Responses to Fungal Biocontrol Agents. In: Ann Rev, Phytopathol. p. 21–43.
Walters DR. 2010. Induced resistance: destined to remain on the sidelines of crop protection? *Phytoparasitica, 38*, 1–4.
Weindling R. 1932. Trichoderma lignorum as a parasite of other soil fungi. *Phytopath, 22*, 837–845.
Weindling R. 1934. Studies on a lethal principle effective in the parasitic action of Trichoderma lignorum on Rhizoctonia solani and other soil fungi. *Phytopath, 24*, 1153–1179.
Yedidia I, Benhamou N, Kapulnik Y, Chet I. 2000. Induction and accumulation of PR proteins activity during early stages of root colonization by the mycoparasite Trichoderma harzianum strain T-203. *Plant Physiol Biochem, 38*, 863–873.
Yedidia I, Srivastva AK, Kapulnik Y, Chet I. 2001. Effect of Trichoderma harzianum on microelement concentrations and increased growth of cucumber plants. *Plant Soil, 235*, 235–242.
Yedidia I, Shoresh M, Kerem Z, Benhamou N, Kapulnik Y, Chet I. 2003. Concomitant induction of systemic resistance to Pseudomonas syringae pv. lachrymans in cucumber by Trichoderma asperellum (T-203) and accumulation of phytoalexins. *Appl Environ Microbiol, 69*, 7343–7353.

Chapter 4
Conflicts of Interest and Commitments

Conflict of interest and commitment are issues that must be managed. The consequences of for universities in the USA, at least, for not properly managing COI and COC are substantial.

This is as it should be. The author began entrepreneurial activities in 1993, as defined as starting companies based on his technologies and playing a role in them. These activities were only loosely monitored and followed by university administrators. This was a relatively uncomfortable position—it was never certain if there were activities that might be misconstrued, and situations where either the author as a Cornell faculty member or the university itself could be liable for unintended consequences of any specific activity. The author therefore was left, in large part, with a considerable amount of freedom but with that freedom was the concern that, if the freedom was misapplied, there could be unpleasant consequences.

This changed recently. The university was required to alter its benign neglect of this topic. This was an unsettling time, because the author feared that there could be a prohibition on the types of activities that he had been pursuing for the past many years. In response to concerns he had, Vice Provost Robert Buhrman provided a thoughtful set of comments on these issues. They are quoted here, because they are germane to this book and to the functioning of any university entrepreneur. From Vice Provost Buhrman

> The day is now past where we can continue to allow our entrepreneurial faculty to pursue their activities with only the barest minimum level of guidance and oversight. Increased federal regulations, and enhanced public scrutiny and skepticism, require that we move from the environment where we could rely on individual faculty to make the appropriate decisions regarding the identification and management of the financial conflicts of interests and the intersecting relationships that are inherent when our faculty and staff engage in entrepreneurial activities, even when this is done in only the best of intentions as in your case. As you know, at Cornell until very recently fC0I (faculty conflict of interest) issues have been handled on a college-by-college basis. As a number of internal and external audits have made very clear, this approach has resulted in inconsistent and inadequate oversight of fC0I, and also of external engagements of faculty when there is a personal financial

G. E. Harman, *Academic Entrepreneurship*,
https://doi.org/10.1007/978-3-031-06821-8_4

interest involved, along with other, not self-interested motives. This has put the institution at considerable risk and the situation has to be remedied, even if there has not been any smoking at Cornell-Ithaca. To address this situation, we have developed, and the Board of Trustees has approved, as you know, annual reporting process for privately disclosing personal financial interests that might be related to the Cornell activities of our faculty and staff, and also of their external professional commitments.

COI Policies and Enforcement Are Essential

There have been egregious examples of COI from a profit motive, and these have been documented in many places, including a report with 52 specific recommendations from the American Association of University Professors (AAUP, 2012). The AAUP has been concerned about both the perception and reality of COI in the university since at least 1915 when it issued its first report, which called attention to risks to higher education from the influence of "commercial practices in large vested interests are involved." The 2012 report goes on to state that the framers of the 1915 declaration could never have considered "a corporation offering a university president offering a university president hundreds of thousands of dollars to serve on a corporate board, or a start-up firm offering faculty members stock options and research funding to test products in which they have a direct financial stake." Further, the AAUP report (page 31) provides six goals that are very important to this discussion and that are highlighted in the following because they are so important. The report states that "a growing body of empirical evidence … shows that inadequate or misguided management of industry relationships threaten the very principles that universities hold most sacred." These principles are listed below, and while they are not highlighted in that report, they will form a basis for this chapter.

1. Academic freedom and independent inquiry: this principle states academics require the right to publish and research autonomously, that is the ability to pursue a project or inquiry without hindrance and to hold and publish opinions on controversial or unpopular topics.
2. Scientific objectivity and research accuracy: this principle states that scientists and others in the academe will report data and conduct research as directed by the data and not by external or internal forces, including outside public pressure for a particular outcome, or by personal bias based on personal gain, expectation of recognition, or academic priorities.
3. Broad dissemination of knowledge: data generated should be provided by academic scientists to the entire world. Data should not be censored by outside funding agencies or by personal biases, and all of the data should be published and not just the portion that is favorable to a funding agency or client or favorable to the scientist's conceptions.
4. Independent analysis and research verification: data should be available to the public and for independent analysis and research verification. This usually includes making biological or other materials available through some mechanism.

5. The development of products that serve the public good: the concept of developing products usually is the province of commercial companies and not public universities or institutions. Thus, while the first five principles apply specifically (though not exclusively) to the academe, the latter one clearly implies a critical role for extension of efforts that proceed beyond providing data and knowledge on to the actual creation of tangible products that frequently have monetary value (the usual definition of products).

These ideals are important and complex and there are many different possible arrangements that can be problematic to universities. The difficulty is to find a path that protects the first four goals and still allows the fifth goal to be compatible with the other four so to fund sources that will allow the freedom and responsibilities that are implied. Even most institutional funding, e.g., by grants is bound by specific objectives so that full and independent inquiry is not fully possible, and of course, many academic jobs have specific duties that preclude this freedom. For example, in agricultural sciences, job duties are frequently defined by research on one commodity and one discipline within that commodity. The first example is Chap. 3 which is an unusual exception to this lack of freedom but one that paid off handsomely for the university and the scientist involved.

Another issue is that university faculty and staff are paid by their institutions and, in the view of some, it is improper for these persons to obtain additional compensation for their activities. Concerns regarding financial reward to entrepreneurial faculty are in part a valid concern about "double dipping" and are an elitist attitude that holds that activities which enrich individual faculty and staff, or through crass commercial activities, are beneath the high ideals of the university. This last statement is worded deliberately to point out an attitude that unfortunately is all too common among faculty and that contributed strongly, but subtly, to the university culture that strongly discourages individual faculty entrepreneurship, even if it provides advantages to the university through royalties, grants, and other activities and to the community and nation that pays the bills for the institution.

However, as pointed out earlier, in many cases the only way that a university faculty researcher can ever see his work developed and used in the real world is to start his or her own company. If that is the case, then the faculty member involved is to have "skin in the game" in the form of investment in the company and stock or options that will provide a major financial advantage to the faculty member if the technology does indeed become useful in society. Taking a project from concept to commercial reality is, for many of us, a primary driver for a company start-up.

Further specific examples will follow in the section on specific types of COI that can occur.

In this chapter, we will briefly examine (a) why universities must have uniform policies for faculty start-ups, (b) what the goal of such policies ought to be and the essentiality that these policies not be viewed as punitive by faculty, (c) specific types of COI that may occur, (d) some specific additions and comments on the 52 AAUP recommendations, and (e) a description of the Cornell system that, if the underlying structure and university culture are transparent and nonconfrontational, is a recommended model. This chapter will not, however, review or examine each of the 52 AAUP recommendations. The reader is referred to the report and, unless specifically commented on, can assume that the author is in agreement with these recommendations.

The AAUP report addresses several issues and, in some cases, egregious abuses of the university system. For example:

1. Should COI considerations and ethics prohibit certain types of research sponsorship? For example, it became apparent that the tobacco industry wanted to fund university research that would whitewash a health hazard and that sought to suppress data that was damaging to the industry's financial well-being.
2. Companies or even nominal nonprofit funding sources may effectively "bribe faculty members to, for example, publish articles with doubtful product claims, dubious financial assessments, or attacks on well-established science" (AAUP, 2012) including those that are politically motivated.

Further specific examples will follow in the section on specific types of COI that can occur.

Ideal Goals of COI Regulations

The goal of any COI policies and regulations ought to, first and foremost, safeguard the ethics and reputations of the university and the individual faculty members. This is not necessarily an easy task in the author's experience; in the absence of clear guidelines, he worked diligently to accomplish the five goals that were set out just above, including the last one, which is the development of products that serve the public good. The last goal required that he be involved with companies and financial institutions that had at least the potential for his acquiring substantial financial benefits. Specific principles that ought to be instituted are as follows:

1. <u>Conflict of interest needs to be managed and cannot be eliminated.</u> As will become clear in the section regarding types of conflict of interest, total avoidance of COI cannot be avoided, and so it must be managed. The policies regarding COI need to be fair to individual faculty and most importantly consistent and clearly communicated to affected persons.
2. <u>Universities need to be transparent to their stakeholders and be able to show strong and uniform policies that appropriately manage COI.</u> Universities receive funding from many sources, including funding from taxpayers through grants, funding from corporations and foundations, and, in the case of public institutions, direct funding from these public institutions. They must be able to respond to inquiries and concerns regarding COI and must actively and continually manage conflicts and concerns. This management has to include consideration of both the perception and the actuality of COI.

3. Universities need to protect their faculty and other staff members. It is possible and even likely for faculty and others to, even with the best of intentions, step over boundaries that others would view as problematic. There needs to be strong institutional guidance; no individual person can easily perceive or understand all of the ramifications of his or her activities. University guidance is necessary to help individuals avoid pitfalls that can potentially damage an individual's reputation and, even potentially, employment or financial situation.
4. The COI policies ought not to be perceived as punitive and should include rewards as well as penalties. University guidelines are usually published and disseminated to individual faculty. They are guidelines in name only and mandate a number of specific practices, such as record keeping and reports, and may impose penalties if they are not met. This really is not different from other university requirements—researchers must publish in reputable journals, teachers must teach effectively, and so forth. There is a difference, however. These other activities also have significant rewards, such as promotion, salary increases, and higher social standing within the institution. Rewards also need to be part of the entrepreneurial structure just as they are for other academic activities.

Types and Categories of COI Situations

All funding ought to be disclosed whenever a publication or presentation is produced. Any audience reading a published work ought to know who funded the research or development. This is an exercise in necessary and obligatory transparency. As published in *Nature* (Editorial, 2012), "Scientists must remember that however irrelevant their involvement in industry might seem to them, others will see it differently—only full disclosure will avert the taint of scandal." Some specifics that pertain to the source of funding follow; some of these were recognized in the AAUP report and some were not.

Research funded by any organization should be reported with the publication of the research. The readers and others need to know whether to discount the work, according to their own biases and perceptions. This ought to always be the case, whether the research is funded, for example, by the National Science Foundation or by company X. In some cases, this notation will add credence to the report if the funding is from a prestigious institution.

This is especially true if the person published the work has an equity position in a company that funds the work. Then there MUST be a statement that "this research was supported by XXXX, in which the author has an equity position." There is an important additional consideration in this case. Some universities do not allow funding of a faculty's program from a company in which the faculty member has an equity position. This prohibition creates a circumstance where the only way, as in the case of the author, where technology can be developed (i.e., to cross the valleys of death described in the previous chapter), is from a company in which he or she has an equity position. This was (and may still be the case) at MIT (Shane, 2004).

However, many faculty there started companies there had access to a strong investment base that was more than happy to support researchers and set them up in a private laboratory. In this case, the investigators took a part-time position at the university and developed their technology in the private laboratory. This is rarely the case with most areas and institutions and certainly is not the case in upstate NY, where funding opportunities from investors are extremely limited. At institutions where there is not a strong investor base, there is no possibility of development unless, either through company grants or direct funding, there is not a provision permitting funding from companies in which the investigator has an equity position. The concept is developed more fully in the next chapter.

Any consulting arrangements that may influence the outcome of report need to be disclosed. In the article mentioned above (Editorial, 2012), the report in question, which involved hydraulic fracturing (fracking) to recover oil, was first considered an independent analysis. However, it was discovered that the author was a well-paid board member who earned more than $400,000 from a company involved in fracking. The *Nature* paper quoted him as saying that his financial stake in the company "would not have served any meaningful purpose relevant to the study." This statement is from an experienced scientist who ought to understand that what he views as an irrelevancy may be perceived as absolutely relevant to someone reading the report. This is precisely why university COI policies must be in place—they can safeguard the reputation of both the institution and the scientist involved. An institutional framework and guideline for COI can prevent situations that are perceived as strongly negative even from faculty and other staff that are motivated by high ideals and themselves seek to avoid the appearance of impropriety.

1. Research that is supported in part by royalties paid to the university, for which the investigator receives a portion of the royalties, or where an invention disclosed to the university may result in payments to either the investigator's research program or paid directly to the investigator needs to be disclosed as well. This usually is an issue for which the university does not view as COI and does not require disclosure on publication.
2. This is a conflict just as much as an investigator who has an equity position in a company. There is usually a difference only in degree, but there really is no difference. In the author's experience, especially early in the development of a technology, the royalties received both to his program and to himself were the primary financial advantages that he gained.
3. All financial remuneration to a faculty member for services, royalties, honoraria, or similar categories needs to be reported annually to the appropriate university office, and this ought to be required by the university.

    1. A full description of the work to be conducted.
    2. Rights to publish freely cannot be compromised (it is the responsibility of the faculty entrepreneur to ensure that timely publication occurs).
    3. All rights and title to IP already owned by the company remains the property of the company. All rights and title to jointly owned IP shall vest jointly in the university and in the sponsor. This usually means that any IP developed

during the course of the university-based research will be the property of the university, although joint ownership may occur as defined by agreement with the technology transfer office.
4. The sponsoring company receives a preferred position in licensing any IP that may result but the company is required to diligently develop and market the IP, otherwise the license to the IP to the company may be revoked.
5. Any liability incurred will be the responsibility of the part incurring this liability.

Transparency Is a Key Element

The fact that these categories involve conflicts are, or should be, entirely legitimate, if full disclosure is made. The sections above addressed several mechanisms to ensure transparency, including full disclosure of all financial support when publications that include any of the situations just described. The development of research contracts between the university and the company, with the academic entrepreneur being the person who conducts the program is a useful avenue of transparency to the university. Again, from the Nature paper, "the key is transparency because this is the basis for trust between institutions and the wider public, which is particularly when people are buffeted by confusing, contradictory and inflammatory. information.... What it [the public] needs is full disclosure." However, as described later in this chapter, there may be additional restrictions subject to the dictates of the Conflict of Interest policy of the university.

Situations Where COI Probably Cannot Be Managed and Must Be Avoided

1. Academic entrepreneurs and others ought not to require students or staff ought not to be specifically for the advantage of the company during the course of their employment at the university. In other words, the university cannot pay for tasks that are not part of their assigned duties at the university. If there are tasks that benefit the university and thc company, then these need to be defined in a contract between the university and the company such as that defined in the research agreement such as that described above.
2. It is never appropriate to delay publication of research that is at the expense of a student, for example, that would preclude the student from publishing his or her research or competing for a job. Such delays may occur for many reasons, such as financial gain of the company including protection of IP or for reasons of academic priority for non-entrepreneurial faculty. Delays that will harm a student are never defensible and, if this situation is likely to occur, the faculty

member ought not to mentor or guide students. For entrepreneurial faculty, a little planning ought to suffice to avoid the problem. IP concepts can be protected through a provisional patent. Provisional patents can be disclosed to the technology transfer office of the university and can be filed with the patent office. These applications are not opened or examined for 12 months, but during this time the potential IP is protected. To complete the protection, a regular utility patent must be filed no later than 1 year of the filing of the provisional. This gives time to complete the invention and still allows students or others the ability to publish without jeopardizing the potential IP.

3. It is never appropriate to unfairly exclude from patent authorship a student post-doctoral fellows or associates authorship. Successful patents generate royalties, and typically a portion of the royalties are distributed back to the inventors. This is the most common intolerable conflict observed by the author and one that is almost never part of the purview of COI committees. In at least one case, the author intervened on behalf of a postdoctoral fellow that had been unfairly treated. This issue is quite easy to deal with at the beginning, when the patent disclosure is filed with the technology transfer office. In any situation where there is any question, the patent disclosure needs to be accompanied by a statement authored by the first author and agreed to be any others, whether at the university or otherwise, that describes the inventive contributions of any and all participants. The technology transfer office ought then to make the determination of appropriate authorship. If this is not done, then conflict and bias, whether perceived or real, are likely to arise. The technology transfer officer does not, or certainly should not, have any conflicts and ought to be able to decide, frequently in conjunction with a patent attorney, these important questions. The time to decide on the authorship is at the time of filing. If this is left to later, there is the potential for serious hard feelings or even lawsuits.
4. In the author's view, it is never appropriate for a full-time faculty member to become the chief operating officer or chief executive officer for a company while he or she has a full-time academic appointment. The requirements of such positions are time-consuming and, even more so, intellectually consuming. Most companies must deal with issues of fundraising, paying the bills, and salaries of employees and the other myriad duties of persons in these positions in any company. If such duties are anticipated by a faculty member, he/she needs to go to reduced time in the university to deal with those company needs. Otherwise, there is a conflict of interest as to who the faculty member is working for and a conflict of commitment as regards time and mental energy available to both the company and the academic institution. However, it is reasonable for the faculty member to function as the chief scientific officer, so long as the goals of the R&D program are (a) disclosed to the university and (b) ideally covered by a contract such as that described above.
5. It is inappropriate for a senior faculty member or administrator to place implicit or explicit pressure on a junior faculty member, student, postdoctoral fellow, or staff to vote or act in ways in accordance with the views of the senior faculty members or administrator. In the cases of votes, paper balloting procedures that

preserve anonymity ought to be used. In other cases, any pressure to conduct in ways that the junior person feels is unethical or morally ambiguous must be avoided.

Other Situations of Conflict of Interest

Conflicts of interest may occur because of excessive timidity on the part of individual faculty and/or by financial pressures to gain sufficient funding to pursue the most appropriate research. The AAUP report cites several examples of these and points out that companies may vigorously defend their interests in preventing or suppressing data not favorable to their positions (AAUP, 2012). Such interference is inimical to the free exchange of ideas and concepts that are part of academic freedoms. In Chap. 3 the concern was expressed that large AgBio companies may be a position to control a large part of the world's food supply and that universities were no longer able to provide an adequate counterweight to this domination of this critical field. This certainly is the case as regards even the ability to choose appropriate projects and fear that if negative data is published that the ability of the project leader to secure funding to conduct appropriate research is at risk. Some examples are as follows:

Beyond that, funding for applied university scientists is so precarious that the large AgBio companies can dictate, in large part, what studies are done or what data is published. An ongoing example concerns the distribution and use of herbicides for application onto transgenic crops resistant to the chemicals. Not surprisingly, this large use of single chemicals across a major part of the food-growing region of the world has resulted in negative impacts, including resistance to weeds. Some weed scientists have reportedly asked if they are "so committed to maintaining profits for the agrochemical industry that we cannot offer up realistic long-term solutions to this pressing problem?" while others have stated that they cannot present data because of fear of their funding sources (http://ecodevoevo.blogspot.com/2013/02/resistance-to-glyphosate-resistant-weeds.html). This perceived problem is surely a conflict of interest if data is behind withheld or research is not done because it may negatively impact potential funding. This entire situation is a reflection, however, of a disparity between the sizes and power of the private vs public research organizations. However apparent conflict on the part of weed scientists largely overcame. The Weed Science Society of America is working diligently to provide solutions to the problem, and it is very widely published, so it is clearly not a universal or systemic issue in this case. However, as public institutions become weaker and weaker and less able to find ways to reinvent themselves, the need for data will become increasingly filled by private concerns who have a strong interest in their own advantage and less in providing a diverse view of their activities or those of agricultural or food issues. This situation is probably more severe than COI generated by timidity on the part of academic scientists may be prudent. The *New Yorker* magazine described the case of Tyrone Hayes, who is a faculty member of the University of California, Berkeley. He was funded by Syngenta to conduct research on the herbicide atrazine, which is used very widely on corn. This family of products was suspected of affecting sexual development in frogs, and he was asked to investigate this possibility. His research indicated that the effect was indeed real and extreme. Of course, if this effect was true that was a high likelihood that it would affect other organisms including humans. Unfortunately, according to the article, Syngenta set out to destroy his scientific credibility. However, this allegation has largely been rebutted, and much of Hayes research has not been reproducible (http://www.forbes.com/sites/jonentine/2014/03/10/de-the-new-yorker-botch-puff-piece-on-frog-scientist-tyrone-hayes-turning-rogue-into-beleaguered-hero/).

This case does illustrate why scientists can feel endangered or prosecuted if they conduct research that may be construed as opposed to the interests of large economic interests, and this can have a chilling effect on choices of research areas. Academic scientists who publish inconvenient results affecting large entrenched economic interests understand that in any contest of wills, they will not be engaged in a fair fight. The propaganda machines, resources, and power of those academic interests far outweigh any defense that an individual scientist can mount.

This is all the more reason why academic institutions need become a more credible counterweight to large public corporations that have very strong academic power. This is an object lesson in why it is inherently dangerous for a few large corporations to largely control major crops like corn and soybeans.

Of course, it is also an object lesson in the relative potential abuses of individual entrepreneurs versus large businesses that can exert so much pressure on academic scientists. It is essential, it seems to the author, that universities and other academic institutions become a more effective and credible counterweight to these large corporations.

Ownership of Inventions by Academic Entrepreneurs

It is essential, and usually required, that faculty direct invention disclosure to the TTO of the university, along with associated agreements, including material transfer agreements and conflict of interest agreements. If a company is involved, they may make agreements, or, if allowed, consulting agreements may be developed by faculty outside the university, but any agreements that involve the university must be submitted to, and executed by, the university. There are very practical reasons for this. First, they involve university property. The IP and research materials developed by faculty or others while in the employment of the university are owned by the university and not by the person who develops these. This is a difficult concept for scientists who naturally assume that the thoughts and concepts developed by them are their personal property, but this is not the case. At least at Cornell, it does not make a difference if a concept occurred at 3 am, while on vacation, or during the normal working hours of the institution, the IP or the research materials that are developed are the property of the university. If a faculty member or scientist is working part-time then the ownership of materials and IP needs to worked out with appropriate administrative persons at the institution prior to actual discovery, product development, or the like. Second, the agreements and IP are complex documents.

Chapter 5 deals with the importance of such documents and the pitfall that can occur if documents are improperly worded and then signed and executed. These may commit the person signing the document and the institution to terms with which they cannot agree. Signing of material transfer agreements may commit materials to two conflicting parties.

The range of unpleasant consequences to the university and to the person who signs the documents are myriad. Any faculty member who understands the potential consequences should be happy that such responsibilities are appropriately the responsibility of others who understand the complexities involved.

The AAUP report includes one recommendation (AAUP, 2012) with which the author absolutely disagrees. The report states that

> Faculty members have fundamental rights to direct and control their own research do not terminate when they make a new invention or other research discovery; these rights properly extend to decisions involving invention management, intellectual property (IP) licensing, commercialization, dissemination and public use.

This "inventor-free agency" is a concept fraught with potential for abuse and strong COI issues. First, the inventor almost always has a financial incentive in the filing of the patent, if inventors negotiate the terms of use, then the perception will be that the inventor is primarily interested in his or her own interests and not those of the university or the taxpayers or others who understand the complexities involved.

First, the inventor almost always has a financial incentive in the filing of the patent, if inventors negotiate the terms of use, then the perception will be that the inventor is primarily interested in his or her own interests and not those of the university or the taxpayers or others who funded the research in the first place. It must be emphasized that it is not the inventor who owns the patent but the university. So, if the inventor negotiates, for example, the terms of the license, then he or she is really negotiating on behalf of the university, and that is blatant conflict of interest. Second, it frequently is not in the inventor's best interest to negotiate their own deal or especially to own the patent. The TTO of the university will almost surely have better knowledge and expertise in patent law, licensing, and the like than does the inventor. It is very much in the best interest of licensing of IP to at least a small company for the university to own the rights to the patent. There are many issues that can arise that the university is more capable of dealing with than either a small company or a faculty member. For example, if a large company decides to violate the patent, to use the technology in ways not approved by agreements or otherwise not abide by ethical, legal behavior, small entities have little recourse.

The anticipated action would be to sue, but if there is a large disparity between the size and financial resources of the violating entity vs the small company of a faculty inventor, this is rarely practical. The larger entity has the resources to conduct lengthy delaying tactics, to hire very accomplished legal talent, and in other ways to make the legal playing field entirely impossible for the smaller one.

However, if the university is the owner of the patent, larger entities that might be tempted to abuse a smaller one will think twice if a university backs the inventor. The ability of the author to "hide behind Ezra Cornell's coattails" has been useful in such situations.

This requires that the inventor and the university, along with any licensee, all work together in a coordinated fashion. The ownership and management of patents by the TTO avoid a large potential of conflict of interest and probably is in the inventor's best interests. A frequent reason for a desire of faculty to maintain their independence and manage or own the IP that they develop is in the author's experience driven in large part by frustration with their TTO. TTOs frequently are slow to act and to take the inventor's concerns seriously.

Moreover, they may have unrealistic expectations of financial returns, either as upfront licensing costs or prohibitive royalty fees. Such avarice on the part of the university is self-defeating and certainly leads to a strong desire of faculty to be free from unfair domination (it should be noted that this is not the author's experience, but he is well aware of situations at other universities where this was the case). It is imperative that the TTO office work closely with the inventor and potential licensees to avoid this unfortunate situation. This leads to frustration on everyone's part and really reflects an unhealthy university culture.

The Cornell Conflict of Interest Policy

As noted at the outset of this chapter, it is essential that universities provide clear and defined guidance as to conflict of interest. This is critical for both the well-being of individual faculty, who might otherwise be unwittingly place him/her in a difficult situation and for the university, which is responsible to the people and who are paying the bills. It is essential that both individual faculty and the university to maintain their integrity and to be able to defend their actions to the community at large. For these reasons COI policies that are fairly applied must exist. Cornell has developed a fair policy that provides a good model for this book. There is a matrix that provides a good model for this book. This matrix defines the degree of conflict and indicates situations where no conflict exists, as shown in Fig. 4.1. The matrix consists of different classification of situations as follows:

Does the commercial entity fund research for the faculty or staff member directly or through a subaward (e.g., from a grant from a SBIR entity). The most conflicted situation is where the entity provides direct funding, the less conflicted from a subaward, and the least conflicted is where no funding exists.

Is there a management role for the Cornell employee?
Is there a significant equity interest by the employee, defined as compensation or an equity position worth more than $10,000?
Does the entity hold a license for IP based on the faculty member's invention?

Funding of research	If entity funds research directly or through a subawrd	Management or employee role	Consulting/fee for service	Significant equity interest	Entity holds license for researcher's invention	Recommendation	Management plan
Entity does not fund research	N/A	Management or employee	No conflict	No conflict	No conflict	No conflict	None
Entity does not fund research	N/A	No conflict	Consulting/fee for service	No conflict	No conflict	No conflict	None
Entity does not fund research							
Entity funds research	N/A	No conflict	No conflict	Equity	Entity holds license	No conflict	None
Entity funds research	Entity funds reseaarcy	N/A		No conflict		Conflict exists	Level 1
Entity funds research	Through sub-award of a government contract		Consulting/fee for service	No		Conflict exists	Level 1
Entity funds research					Entity holds license	Conflict exists	Level 1

Fig. 4.1 The Cornell matrix for COI

If it determined that a conflict exists, then a management plan to deal with the conflict may be required. The matrix in Fig. 4.1 is not endorsed by the author but the fact that a well-defined policy exists is strongly endorsed. Academic entrepreneurs must have clear guidelines and rules, otherwise there is great difficulty in knowing what is and what is not allowed. Such clear guidelines are welcomed to by the author, although the implementation of them was nearly impossible for him. At Cornell, there is a set of forms that faculty are required to complete and that determined the nature of any arrangement with a commercial entity.

The factors in combination determine whether there is a conflict or not and impost different levels of management for the program involved. If a conflict exists, then a specific management plan must be implemented, and the faculty member must agree to abide by the terms of the agreement. If the faculty member does not agree or does not abide by the terms of the agreement, then the university could take further steps up to and including termination of employment.

The highest level of conflict management occurs if the faculty member receives direct funding from the entity, but lesser levels of stringency are required if the funding is through a subaward of a government agency, such as SBIR or STTR. The author was required to abide by a management plan at the highest level. He received direct funding from Advanced Biological Marketing in which had a role and an equity interest but received no salary. In the author's case, the requirements were as follows (the terms have been generalized, but they were more specific and list specific grants).

1. He was required to disclose in writing to students, fellows, trainees, and staff who he supervised, collaborators, and publications and conferences his external financial interests in the company. His department chair would ascertain that this requirement is met.
2. If a grant was received by one of the companies in which he had an equity interest was funded, a co-pi would be required, and the co-pi would be required to

attest to progress and avoidance of undue COI during the performance of the funded research.

3. If the author was chair of a special committee for which any graduate student that will be working on company-funded projects, the conflict manager (CM, in this case the department chair) would meet with the committee to ensure that they are aware of the circumstances, to allow them to be aware of this responsibility to take adequate measures to ensure the student's progress, ability to publish and meet degree requirements.
4. The CM would meet with the author on a periodic basis, not less than annually, to review student progress and assess compliance with the terms of the management plant. The CM was required to fCOI committee her findings regarding the results of this review, according to a specific form that was provided.
5. The author was required to provide the CM with an updated list of any students, staff, and postdoctoral students working company-funded projects, and any other persons would be provided with specific information regarding his positions in companies with which he had a funding or equity interest.
6. Persons working at Cornell may not have any part-time employment with the companies with which funding is received.
7. The author could not supervise any employees of outside companies from which he obtained funding.

The author had no issues with the terms detailed above, with the exception of point 6. They all conformed to his standard operating procedures, all that was added from his perspective is a more formal reporting relationship with the CM. He ordinarily provided such informal, albeit in a less formal arrangement.

Point 6 did require a significant modification in his relationship with his staff. For a number of years, two of his long-term employees had approximately half-time appointments at Cornell and half-time appointments with the companies. In their company time, they performed quality control determinations and assisted in manufacturing operation. Importantly, the company maintained separate laboratories and other workspace for these two employees. This was an arrangement that worked well from the perspective of employees, the author, and the companies and assisted in rapid progression of research projects. This was particularly important for the scale-up of potential products and inventions for which Cornell received royalties—it is obvious that the persons who did the original day-to-day project development would be the most efficient at taking products from the lab bench to larger-scale commercial development. It seemed to the author that this was the optimal path to generate royalty revenue to Cornell and bring products to the marketplace for the benefit of society. However, the COI did not agree and would not permit this expeditious arrangement. Their concern, even though separate laboratory and production facilities were provided by the companies, that this arrangement could be "perceived as enabling a situation in which Cornell resources might be used to benefit the companies."

One of the author's guidelines all along is that if Cornell says that a particular practice is not allowed, then that is the final word, regardless of his opinion. This was the case in this circumstance and the author signed the agreement binding him to the terms not above.

Transparency and Implementation of COI Rules

Thus, there was substantial agreement between the author and the fCOI and he agreed to the terms of that dictated after an appeal of point 6. However, the implementation of the rules and the terms of the agreement were poorly handled. The terms of the fCOI were transmitted through an associate dean and the chair. Since this was all in writing, this was not a large issue. However, any response from the author back to the committee had to again pass through the department chair to the associate dean, and the messages the fCOI received were invariably garbled. The author was not permitted to talk with or have any direct communication with the fCOI. As a consequence, month passed without adequate resolution of what was a simple matter that could have been resolved with a short conversation since the committee and the author were in substantial agreement. If however no direct conversation occurs, then it becomes almost impossible to make complex arrangements without misunderstandings. In fact, the author received a letter stating that the fCOI was disappointed that he needed an extension to comply with the terms of the agreement, when, in fact, he already was in compliance. Moreover, one Sunday evening the author received a call at home from the associate dean apologizing because he had forgotten to say that the fCOI and the author agreed.

This was a hugely disruptive set of events that unnecessarily went on for several months. In the meantime, staff jobs and the research program were in jeopardy. This was totally unnecessary since there was no fundamental disagreement between the fCOI and the author, but there was an almost total breakdown of communication.

An absolutely fundamental requirement of implementation of any COI plan, therefore, is that there be an efficient, transparent, and direct communication link between the affected faculty member and the fCOI. The fCOI must be available for discussions with the affected faculty member. Such arrangements are complex and unnecessary insertion of other parties communicating information to and from the faculty member who are not experienced in the events and detail of entrepreneurial activities and should not intervene in specific cases involving affected faculty members. A short meeting between the faculty member and fCOI ought to suffice to erase any confusion and to arrive at a reasonable decision for all affected (administrators at any level from the Dean's or Chair's office ought to be invited so that everyone understands what the rules and regulations are).

As it was, the author was in limbo regarding all of this for many months. Not only was he inconvenienced but his staff's employment and his program were in jeopardy. Again, this did not occur because of any insurmountable disagreement between him and a breakdown of communication.

Finally, the author decided, for the sake of his staff and his research program, he had to retire from Cornell. After that he became a full-time employee at Advanced Biological Marketing (ABM) and became the Chief Scientific Officer. This has worked out very well for all concerned, but the circumstances driving this decision were regrettable and avoidable. These events occurred with no malice or animosity on anyone's part but were a consequence of the opaque barrier between the author and fCOI.

After retirement, the author has received every courtesy and the cooperation of Cornell. ABM initially leased his former lab and office space at Cornell while ABM was building a state-of-the-art research office a short distance away from his former Cornell lab.

Academic Entrepreneurs Need Rewards as Well as Roadblocks and Difficulties

The events just described were unpleasant and unnecessary. However, other aspects of life at Cornell were rewarding and some programs have been welcoming and helpful. For example, the Cornell Biotechnology Program has funded his efforts for many years. They have a mandate for economic development from New York State, and they understand what kinds of structures and rewards need to be in place for entrepreneurial activities, including royalty structures and opportunities for employment from technological advances.

Similarly, the Cornell patent office has been quite helpful, and the author has a good relationship with them. Other parts of the university and appreciate less well what is required and what the benefits of entrepreneurial activities are.

From the information just presented, it is clear that there are major difficulties and disadvantages to an entrepreneurial lifestyle, but there ought to be institutional rewards as well. Such a reward structure will be more fully defined in Chap. 5 and is to the benefit of both the faculty member and the university.

University faculty have numerous requirements, and the classic standard of performance is how well they perform their assigned duties. Promotion, salary increases, and other rewards are based on more or less objective measures of performance. For research, number and quality of publications, international standing based on recognition and awards, numbers and quality of visiting scientists, and invitations to present at scientific meetings are typical measures. For teaching, student ratings of courses, students mentored, popularity of courses, and success of students are objective measures.

It is possible to have similar objective measures for successful entrepreneurial activities. Metrics include patents licensed, royalties to the university obtained and numbers of jobs created, and economic benefits obtained. However, so far as the author is aware, these are not measured or requested for faculty programs even though they are easily measured. Moreover, these need to be available to their stakeholders, including legislators, foundation donors, and the like. Chapter 3 deals with the university culture, how it can be assessed by potential academic and outside entrepreneurs, and way that the universities can become more responsive to both needs and current political realities.

Lessons from Chap. 4

1. Formal management of conflicts of interest must be a part of university management. This is usually managed by a faculty conflict of interest (fCC01) committee that establishes specific guidelines for COI in the university.
2. COI is likely to exist in many faculty programs, even sometimes in ways not recognized by university policy.
3. COI cannot and should not be eliminated in the university system; instead it must be managed in ways acceptable to the university and by the faculty affected.
4. A formal COI policy ought to be welcomed by entrepreneurial faculty and staff. Otherwise, the affected individuals are without rules or guidance; from the author's experience, this is an uncomfortable position.
5. There are degrees of COI and these need to be recognized by the fCOI and requirements for their management clearly communicated to the affected faculty member. Policies for each individual program are likely to differ, but these must be in accordance with well-publicized policies. It must be understood that while these are designated guidelines, they are in fact absolute requirements.
6. COI rules frequently are viewed as punitive, but in reality they are no different in principle from other rules regarding performance, including objective criteria for judging excellence in research, teaching, or other academic pursuit.
7. It is imperative, however, that communications between affected faculty and the fCOI be handled in a transparent manner. Affected faculty must be able to directly communicate with the fC0I; otherwise, ill feeling and potential significant harm to the affected faculty member is highly likely.
8. Success in entrepreneurial activities ought to be rewarded in terms of faculty promotion, prestige, and other academic value. There are objective measurements of success in entrepreneurial activities and these are discussed in detail in the next chapter.

References

AAUP. (2012) *AAUP recommended principles & practices to guide academy-industry relationships*. pp. 307.

Editorial, N. (2012). Unfortunate oversight. *Nature, 488*, 5. https://doi.org/10.1038/488005a

Shane, S. (2004). *Academic entrepreneurship: University spinoffs and wealth creation*. Edward Elgar.

Chapter 5
Formation of Companies from an Academic Base

The first step in formation of a company is to create a legal entity. One of the most useful is a limited liability company. To do this obtain, a copy of the article of organization from the state where you want to organize the company, and then choose a name for the company. Fill out the LLC artless organization form, publish a notice in your local newspaper, and submit the articles of organization to the state where you want to apply. The advantage of an LLC is that it limits your personal liability or, in other words, protects your personal assets. You will also establish the officers of your organization, decide on the composition of the Board of Directors, and decide who your auditor will be. Other types of companies are S or C corporations. These differ in that a C corporation is taxed separately from its owners, while an S corporation is not taxed separately from its owners.

In most cases, a new start-up will require a business plan. This is an important document that provides the blueprint for all of the new company's activities. Important components of the plan include what the vision of the company is, what products will be, how will the products be manufactured, how will the necessary funding be obtained, and what financial projections can be made. The financial projections need to be realistic, otherwise the creditability of the company and its founders will be jeopardized. This is a primary document that will have to be provided to prospective new stock buyers or other investors.

Team building is a critical first step. No entrepreneurial activity functions as a single person (although it may be useful to have a single owner, especially for consulting). So, teams are very important. Thc founders are the initial team members, but others are needed. Successful entrepreneurs need a collaborative team, frequently with members both within and outside the academe. The right team can identify technologies, bring it to commercial life, steer it through challenges, and create a solid company (23). The Board of Directors ought to include persons knowledgeable in the field, who are respected, and who can bring in the necessary financing. You may need persons with capabilities in marketing. Creating a congenial but challenging team is one of the most critical aspects of any project or

G. E. Harman, *Academic Entrepreneurship*,
https://doi.org/10.1007/978-3-031-06821-8_5

organization. The project will begin with the inventor/founder of the project. Financing may be possible through loans. Unfortunately, while the loan will be to the company, they have to be signed by an officer of the company. This results in a personal liability to the signee, so if the company is unable to pay, the signee must. This is a risk to the signee. The author was never willing to undertake this obligation. However, others do. For example, Dan Custis of Advanced Biological Marketing (see Chap. 6) largely financed the company in its early years through loans—he even twice mortgaged his house. This has resulted in a big payday but that was by no means certain at the outset.

If the project is noncommercial and adaptive and is to be used entirely within the university, the primary team members are likely to be other academics or cooperators outside the university that bring the required intellectual capability to the project. Beyond this, it is likely that there will need to be a wide range of cooperator types, for example, valuable persons are those in countries where products are used and provide expertise on the specific produce being sold.

However, if the project's products are disruptive, it means that commercial entity must be involved, either through licensing to an existing company or through establishment of a new company, and then the issues become more complex. First, what role does the originating scientist want? Does he/she want to control what happens to the company and technology, or will a supportive role suffice? Is an element of control important to him/her or does this person want to the chief executive offer? If so, then a full-time appointment probably is inappropriate since the COIs are too great to continue as a full-time academic. In that case, is she/he willing to move to a part-time academic appointment of move fully into the company? One role that is probably acceptable from a COI point of view is that of chief scientific or technical officer. In that role she/he can work toward the goals of the company and can have a seat on the Board of Directors. In the event that the founding scientist does not become the CEO, then someone has to be selected to fill that position. It is important for a fledging entrepreneur to understand that training as an academic does not provide a good set of skills to manage a company. The CEO may have to be recruited by the originating academic or, better yet, by the BOD.

One critical part of company formation is fundraising. Most academic companies are pre-seed stage and therefore have to raise early funding. This can be difficult, since as described in Chap. 2 revenues are likely to be minimal in the young company. The only asset is likely to the IP or the personal credibility of the founders. IP is expensive and this is why it is so important to have the support of university patent office. Universities are likely to pay for the IP with the expectation of royalty streams in the future.

As mentioned earlier, it is very important to have BOD members that are themselves able to provide funds and to recruit other wealthy individuals to provide funds through sale of stock. Any potential BOD members need to be asked if they are willing to fill this function. If they are not, then they should not have BOD seat. In the author's experience, we had a very well qualified BOD member, but he was unwilling to approach any of his friends to raise the necessary capital. It is critical to ascertain at the outset whether each of the team members are willing to raise the money required for the company to succeed. If there is no capability of raising the funds required, then the company should not proceed.

It will be important for any new product to establish the path to success. What will the products do, and what will they do—it is not enough for the products to function, but they must also provide a profit margin for the users. What is the market size, and how does one reach the marketplace? If the product is disruptive, if there is an understanding in the marketplace that there is a need for the product being envisioned? How long will it take to cross the valleys of death and what are the milestones of progress along the way? Who will be approached for funding, and how will this be accomplished? Who in the team will make the contact? These are all critical steps of planning as defined in the company's business plant. If outside funding is obtained, how much dilution of initial founder equity is acceptable?

Another source of funding is grants. These may come from national or state agencies or private foundations. Some universities or other organizations provide funds based on a competition, in which they select fledgling companies to fund. A very useful US possibility is the Small Business Innovation Research program. This agency funds translational research but does not provide funding for other activities such as marketing or other necessary company functions. Grants from this source have been very useful for the author. One requirement of SBIR funding is that after preliminary approval of the grant, an audit is conducted. If the company is not a going concern, i.e., able to be financially stable, then the grant will not be approved.

Chapter 6
Agreements, Contracts, Regulatory Affairs, and Royalties

Entrepreneurs must deal with many types of legal documents. An important part of this is that unless you are acting in private capacity, you will not be able to sign these documents. Signatory authority is restricted to officers of the organization. Consequences of improper signing can include lawsuits or termination of employment. I have known cases of improper signing and it took years of discomfort to resolve.

Contracts are only of value if they are enforced. In cases of disagreement, the only recourse is a lawsuit regarding the matter of dispute. Small entities frequently lack the resources to do so. In my experience, when we told a company they were in violation of the contract, they said "so sue us," knowing that we lacked the financial resources to do so. An important adage is "he who has the gold rules."

Confidentiality and Material Transfer Agreements

A confidentiality agreement is a legal agreement that binds one or more parties to nondisclosure of confidential information.

These agreements are likely to have the following components (Example 6.1 contains language of such agreements):

- The parties involved and the dates of the agreements
- Obligations of each party
- Exceptions to the agreement, such as prior disclosure in a publicly available document
- All parties agree that only documents provided in writing shall be deemed confidential
- Signature pages
- Confidentiality agreements are often combined with material transfer agreements (Mtas). These include:

G. E. Harman, *Academic Entrepreneurship*,
https://doi.org/10.1007/978-3-031-06821-8_6

- Uses to which the material can be put
- Provisions requiring destruction of materials at the end of the agreement
- Prohibition of commercial sales of the material without agreement of the providing party
- Any royalties to be paid by the receiving party

Publication Agreements

All publishers require publication agreements. The contract for this book is provided in Example 6.2. Typical provisions include:

- The rights and responsibilities of each party
- Any remuneration to the author including royalties
- Publisher's responsibilities
- Rights of the authors to publish or reuse the published material
- Author's responsibility to deliver the manuscript
- Copyright restrictions
- Ethics rules for authors

Contracts

There are many types of contracts. Generally, they are used whenever two or more parties wish to be legally bound for any purpose. Examples include:

- Deeds
- Any of the agreements mentioned above
- Purchase agreements
- Sales agreements

- One type of agreement is grants. Almost everyone is familiar with these but included is a statement such as "the grantee receives worldwide exclusive rights to the invention upon x% of sales of the product."

Regulatory Affairs

Many governmental agencies require regulations. These include:

- The Department of Transportation
- The Environmental Protection Agency
  - Food, Insecticide, Fungicide, and Rodenticide Act
  - Air quality
  - Environmental protection

- The Food and Drug Administration
- The Department of Health and Human Services
- The Mine Safety and Health Administration
- Motor Vehicle Offices

In most cases, regulations are concerned with safety and effectiveness. They require statistically analyzed data and there may be multiple tiers of trials. They frequently set levels of allowable concentrations of materials. An example of a pesticide registration is provided in Example 6.3.

Royalties

Royalties are agreements that provide, typically, a percentage of net sales to one or more parties. They can be valuable to academic entrepreneurs since they provide direct payments to the originating parties. At the University of Wisconsin, the inventor or inventor group receives 20–40% of the gross royalty income generated (https://research.wisc.edu/intellectual-property/royalty-income-sharing/). At Cornell, inventors are entitled to XXXX.

Patents

Patents provide exclusive use of a product or service for 20 years from the first date of filing. It should be noted that patents are a negative right—it prevents others from using the patented information, but it does not necessarily mean that you can practice your invention. It may be that there is a broader invention that prevents any commercial use of the invention.

Typical sections of patents follow:

- Abstract
- Related patents or other publications.
- Field of the invention
- Background
- Scope of the invention
- Examples
- Claims

The field to the invention describes the general class into which the invention falls.

Examples contain data and descriptions of the subject. They provide the basis for the claims that follow.

Claims describe the particular item being patented. Frequently, claims deal with the composition of the matter being described or methods of use. They cannot be indefinite and so must claim an effective range of the material being patented. For example, a claim of x – y. It is important to claim as wide a range as possible; in some cases there can be claims of x – y and 2x – 2y. Frequently there are claims followed by subclaims that are more specific and limited than the original claim.

Nearly every patent is rejected at first. This does not mean that the patent will not issue, but only those conversations need to occur between the patent attorney and the patentee. Even final rejections may be appealed, usually in face-to-face meetings. Example 6.4 gives the text of an issued patent.

Example 6.1: Confidentiality and Material Transfer Agreement
Confidential

MATERIAL TRANSFER AND CONFIDENTIALITY AGREEMENT ORGANIZATION X-ORGANIZATION Y

This **Material Transfer Agreement**, (the "**Agreement**") is made effective this _____ (date)_____ and is by and between Organization X, address and Organization Y, address, (hereinafter "**Recipient**").

WHEREAS, the parties are engaging in discussions and material and information transfer potentially leading to a business arrangement between them relating to the assessment of Biological Seed and Other Treatments for Enhancing Crop Performance and Yield,

WHEREAS, the Biological Agents and Formulations are proprietary to Organization X,

WHEREAS, in the course of the Discussions the parties have determined that Organization X will provide certain materials for testing and,

WHEREAS, the parties wish to have an agreement governing the treatment of such materials, testing and information generated thereby;

NOW THEREFORE, the parties agree as follows:

1. **Confidentiality Agreement.** Confidential material may be provided by either party to the other. Any confidential materials shall be in writing and clearly labeled as confidential. Upon understanding the nature of the material, the receiving party can refuse to accept the Confidential Information.

   a. **Recipient** will not disclose the source or nature of the materials provided by Organization X to any third party without the express written permission of Organization X.
   b. Confidential information does not include:

      i. Information that is now in the public domain or which subsequently enters the public domain through no fault of the **Recipient** or in breach of this Agreement
      ii. Information received from any third party having a lawful right to make such a disclosure;

iii. Information which can be demonstrated by written records to have been known to the **Recipient** prior to its disclosure by Organization X information for which the disclosing Party has waived confidentiality in writing; or
iv. materials or information that is independently developed by the **Recipient** without use directly or indirectly of the Confidential Information received from Organization X provided that such independent development can be substantiated by written records and documents, in which case the **Recipient** shall give immediate written notice to **ORGANIZATION X** with the detailed description of the basis for challenging the status of the disclosed information as Confidential.

2. **Materials and Testing. ORGANIZATION X** will provide formulated test materials as coded product names and provide directions for their use.

   a. The products provided by **ORGANIZATION X** will be used by **Recipient** only for seed or other agreed upon treatments and will be used only for the purposes of evaluation of plant performance in **Recipient's** trials.
   b. The microbial strains and formulation compositions are proprietary and the property of **ORGANIZATION X. ORGANIZATION X** has, or is engaged in developing, genetic and biochemical tags for all of the strains that it will provide to **Recipient**.
   c. **Recipient** will not attempt to isolate or identify any of the microbial strains provided by **ORGANIZATION X** or allow others to do so.

3. All materials provided to **Recipient** shall remain the property of **ORGANIZATION X. ORGANIZATION X**, at its sole discretion may choose to patent, register with EPA, state or other regulatory agencies, or the Organic Materials Review Institute any of its strains, uses, formulations or products; any such patents or patent applications or registrations shall remain the property of **ORGANIZATION X**. However, if a subsequent agreement for exclusive use by **Recipient** of any product, formulation or strain, is agreed between the parties, then such exclusivity would include the rights to use the patents, patent applications or registrations royalty free.
4. **ORGANIZATION X** shall be the exclusive supplier of its proprietary products to **Recipient**.
5. **Research Report**. **Recipient** will provide **ORGANIZATION X** a written research report of all tests conducted, including methodology, soil types, weather, locations and analysis of data (the "**Research Report**") within 90 days after completing the Testing.

   a. **Ownership and Use of Testing Results.** Any reports from **Recipient** shall be deemed Confidential Information of **ORGANIZATION X**.
   b. **Limitation of Use; No Reverse Engineering.** The Materials will be used only for the Testing. The Materials will not be used for commercial or other purposes. **Recipient** shall not analyze, or allow to be analyzed, the Materials to determine its biological formulation, chemical composition, or structure

or ingredients. **Recipient** shall not isolate or attempt to isolate any microorganisms in the materials provided by ORGANIZATION X, nor shall **Recipient** allow others to do so. **Recipient** will not engage in any effort to reverse engineer, duplicate and circumvent **ORGANIZATION X**'s technologies. **Recipient** shall not attempt to isolate, reproduce, modify, or propagate the Material.

c. **Control of Materials**. **Recipient** shall retain control over the Materials and not transfer the Materials to any person or entity for any purpose other than field or greenhouse testing. At the end of the testing season, all biological materials, including materials provided by **ORGANIZATION X** and seeds treated with its products, shall either be destroyed or returned to **ORGANIZATION X**. If materials need to be retained beyond the testing, then approval shall be sought in writing from **ORGANIZATION X**.

d. **No Warranty**. **THE MATERIALS ARE BEING SUPPLIED TO RECIPIENT "AS IS", WITH NO WARRANTIES, EXPRESS OR IMPLIED, AND ORGANIZATION X EXPRESSLY DISCLAIMS ANY WARRANTY OF MERCHANTABILITY OR FITNESS FOR A PARTICULAR PURPOSE OF THE MATERIAL, HOW IT INTERACTS WITH THE OTHER PARTIES' MATERIAL OR EFFECT UPON SEED OR PLANT GROWTH.**

e. **IP Warranty. ORGANIZATION X** warrants that it (i) has the right to provide the Materials hereunder and (ii) has no reason to believe that its Material, or the use of the Materials on seed infringes any patent or other proprietary or intellectual property right of a third party.

f. **Hold Harmless. Recipient** shall indemnify Provider and hold Provider harmless from any loss, claim or liability of any kind which arises as a result of Provider's or **Recipient**'s use, handling or storage of the Materials.

g. **Compliance with Laws**. **Recipient** will comply with all laws, regulations, and/or guidelines applying to the use of the Material and assume sole responsibility for any claims or liabilities that may arise from or as a result of **Recipient**'s use of the Material.

6. **Binding obligations.** Each party specifically agrees and warrants:

a. that the obligations and commitments on it as a **Recipien**t shall be binding upon each of its employees, advisors or consultants who receive or have access to the Confidential Information of the Provider; and

b. **Recipient** shall do all things necessary in order to so bind each such employee, advisor or consultant in writing, including acknowledging receipt of a copy of this Agreement and undertaking in writing to be bound; and

c. **Recipient** shall take all action, legal or otherwise, to the extent necessary to cause those individuals to comply with the terms and conditions of this Agreement and thereby prevent disclosure or improper use of the Confidential Information.

7. **Term.** The obligations imposed upon the **Recipient** shall last as long as the Materials are in the **Recipient**'s possession.
8. **Agreement or Obligation as to Business Transaction.** The parties specifically agree:
   a. No agreement regarding a business transaction which is the subject of the Discussions shall be final and/or legally binding until such agreement is embodied in a final, written document, executed by both parties;
   b. Either party may, in its sole discretion, end the Discussions or the Testing at any time.
9. **Intellectual Property**. No intellectual property right or license is granted by this Agreement. All of **ORGANIZATION X**'s intellectual property rights to the Materials in existence or development prior to this Agreement, including without limitations issued patents, filed patent applications or trade secrets, will remain the property of **ORGANIZATION X**.
10. **Remedies. Recipient** agrees to indemnify and hold the Provider harmless from any damages, loss, cost, or liability (including legal fees and the cost of enforcing this indemnity) arising out of or resulting from any unauthorized use or disclosure by the **Recipient** of the Materials or other violation of this Agreement. In addition, because an award of money damages would be inadequate for any breach of this Agreement by the **Recipient** and any such breach would cause **ORGANIZATION X** irreparable harm, **Recipient** also agrees that, in the event of any breach or threatened breach of this Agreement, **ORGANIZATION X** will also be entitled, without the requirement of posting a bond or other security, to equitable relief, including injunctive relief and specific performance. Such remedies will not be the exclusive remedies for any breach Agreement but will be in addition to all other remedies available at law or equity to **ORGANIZATION X**.
11. **No Waiver.** No failure or delay by any party in exercising any right provided by this Agreement shall operate as a waiver thereof, nor shall any single or partial exercise of such a right preclude any other or further exercise of such right.
12. **Entire Agreement; Modifications**. This Agreement constitutes the entire agreement of the parties with respect to the Discussion and obligations regarding Confidential Information and shall be amended or modified only by an agreement in writing signed by both parties that states that it is an amendment or modification to this Agreement.
13. **Governing Law.** This Agreement shall be governed by the internal law of the State of Ohio

Organization X
Name & Signature

Organization Y
Name & Signature

Example 6.2: Patents

I will provide two links to two patents and provide information on claim language on these.

The first is

Harman GE, Stasz TE, Weeden NF, inventors; Cornell Research Foundation, assignee. Fused Biocontrol Agents. USA patent 5,260,213. 1993. https://number.academy/5165928

The claim language is:

1. A biologically pure Trichoderma strain selected from the group consisting of 1295-7 (ATCC 20846), 1295-74 (ATCC 20848), 1295-22 (ATCC 20847) and 1295-106 (ATCC 20873)
2. The biologically pure Trichoderma strain as in claim 1 selected from the group consisting of 1295-7, 1295-74 and 1295-22
3. The biologically pure Trichoderma strain as in claim 1 which is 1295-22
4. The biologically pure Trichoderma strain as in claim 1 which is 1295-1. A biologically pure Trichoderma strain selected from the group consisting of 1295-7 (ATCC 20846), 1295-74 (ATCC 20848), 1295-22 (ATCC 20847) and 1295-106 (ATCC 20873)

In this case, what is patented is the strain itself—no other use of these stains is allowed, without the approval of the licensee, which is Cornell.

Compare this with the following patent: Smith VL, Wilcox W, Harman GE, inventors; Cornell University, assignee. US Patent 5,165,928. Biological control of *Phytophthora* by *Gliocladium*. USA1992. https://patents.google.com/patent/US5165928A/en?oq=5165928.

The claims in this patent are as follows:

What is claimed is:

1. A method of controlling plant diseases incited by *Phytophthora* sojae Kaufmann and Gerdemann which comprises applying to the plant root biosphere of the plants to protected a biosphere colonizing amount of a biocontrol agent selected from the group consisting of *Gliocladium virens*, 031 (ATCC 20903*Gliocladium virens*, 035 (ATCC 20904), and *Gliocladium virens*, 041 (ATCC 20906).
2. The method of claim 1 wherein the plant disease incited by *Phytophthora sojae* Kaufmann and Gerdemann is stem and root rot in soybean plants.
3. The method of claim 2 wherein the biocontrol agent is *Gliocladium virens*, 031 (ATCC 20903).
4. The method of claim 2 wherein the biocontrol agent is *Gliocladium virens*, 035 (ATCC 20904).
5. The method of claim 2 wherein the biocontrol agent is *Gliocladium virens*, 041 (ATCC 20906).

The claims are quite different from this first case; rather than claiming the strain, this patent claims a method of use. Therefore, stains in the second patent can be used without patent infringement for any purpose other than the method claimed.

Example 6.3: EPA Registration

The registration can be found at https://www3.epa.gov/pesticides/chem_search/ppls/068539-00004-20210426.pdf.

The registration includes statements on safety, crops on which it can be used, and the formulation and ingredients. Importantly, this strain has approval for exemption from tolerance. This allows use on crops without the requirement for testing to determine whether the levels of the fungus are within acceptable limits. This exemption can be found at § 180.1102 *Trichoderma harzianum* KRL-AG2 (ATCC #20847) strain T-22, exemption from requirement of a tolerance.

Registrations are important for commercial use, since without them the organisms cannot be used for control of pests or diseases. In addition, they provide credibility to the product. The registration was a major reason for the successful marketing of T22.

MIX
Papier aus verantwortungsvollen Quellen
Paper from responsible sources
FSC® C105338

FSC
www.fsc.org

If you have any concerns about our products,
you can contact us on
ProductSafety@springernature.com

In case Publisher is established outside the EU,
the EU authorized representative is:
Springer Nature Customer Service Center GmbH
Europaplatz 3, 69115 Heidelberg, Germany

Printed by Libri Plureos GmbH
in Hamburg, Germany